Advances in Insect Physiology

Volume 35

Advances in Insect Physiology

edited by

S. J. Simpson

School of Biological Sciences, The University of Sydney, Sydney, Australia

Volume 35

ELSEVIER

Amsterdam • Boston • Heidelberg • London • New York • Oxford
Paris • San Diego • San Francisco • Singapore • Sydney • Tokyo
Academic Press is an imprint of Elsevier

ACADEMIC PRESS

Academic Press is an imprint of Elsevier
32 Jamestown Road, London NW1 7BY, UK
Radarweg 29, PO Box 211, 1000 AE Amsterdam, The Netherlands
Linacre House, Jordan Hill, Oxford OX2 8DP, UK
30 Corporate Drive, Suite 400, Burlington, MA 01803, USA
525 B Street, Suite 1900, San Diego, CA 92101-4495, USA

First edition 2008

ISBN: 978-0-12-374329-9
ISSN: 0065-2806

For information on all Academic Press publications
visit our website at elsevierdirect.com

Printed and bound in Great Britain by
CPI Antony Rowe, Chippenham and Eastbourne

Transferred to Digital Printing, 2010

Contents

Contributors

J.-L. Deneubourg

Unit of Social Ecology, Université Libre de Bruxelles, Campus de la Plaine, Brussels, Belgium

C. Detrain

Unit of Social Ecology, Université Libre de Bruxelles, Campus de la Plaine, Brussels, Belgium

M. O'Donnell

Department of Biology, McMaster University, Hamilton, Ontario, L8S 4K1, Canada

Insect Excretory Mechanisms

Michael O'Donnell

Department of Biology, McMaster University, Hamilton, Ontario, L8S 4K1, Canada

ADVANCES IN INSECT PHYSIOLOGY VOL. 35
ISBN 978-0-12-374329-9
DOI: 10.1016/S0065-2806(08)00001-5

1 Introduction

Excretion refers to the processes which remove from the metabolic pool substances which interfere with metabolism. In this chapter I will consider three separate aspects of excretion, as first proposed by Maddrell (1971): (1) removal of molecules which are undesirable or perhaps even poisonous at all except very low concentrations; (2) excretion of molecules which are not toxic but merely useless, and which would become obstructive if allowed to accumulate and (3) excretion of molecules which are useful or essential but are present to excess. This would include excretion of water and physiological ions in some circumstances.

In writing this chapter I have attempted to link recent advances in our understanding of the physiology, biochemistry, genetics, molecular biology and chemical ecology of insect excretory mechanisms to the fundamental principles of design established more than 30 years ago. For example, the primary ATP-dependent transporter in insect epithelia was referred to in the literature until the early 1990s as the 'electrogenic alkali cation pump'. This pump is now known to reflect the integrated actions of two separate transporters, an electrogenic vacuolar-type H^+-ATPase and an exchanger of Na^+ or K^+ and H^+ (Maddrell and O'Donnell, 1992; Wieczorek, 1992; Rheault *et al.*, 2007). Similarly, the high osmotic permeability of insect epithelial cells noted in early studies (O'Donnell *et al.*, 1982) is now seen to be based on aquaporins (Echevarria *et al.*, 2001; Kaufmann *et al.*, 2005). A third example is the multialkaloid transporter responsible for elimination of alkaloids such as nicotine, morphine and atropine by insect excretory organs (Maddrell and Gardiner, 1976). Recent studies suggest that this transporter is a P-glycoprotein (P-gp)-like transporter (Gaertner *et al.*, 1998; Leader and O'Donnell, 2005) produced by expression of multidrug resistance (MDR) genes (Tapadia and Lakhotia, 2005).

A recurrent theme of this review is that studies of insect excretory mechanisms frequently provide illustrations of the Krogh principle: "For a large number of problems there will be some animal of choice or a few such animals on which it can be most conveniently studied." In many instances, these examples of the Krogh principle are a consequence of unusual diets, such as blood, or life in extreme environments, as is the case for the tenebrionid beetles of the Namibian desert or mosquito larvae found in saline lakes containing very high levels of magnesium and sulfate.

Adaptations to extreme environments, coupled with the large number of unusual food sources exploited by insects and their extraordinary diversity, provides a fertile ground for research into excretory mechanisms.

Several reviews in the last decade have analyzed literature relevant to this work. Comprehensive studies of insect diuretic and antidiuretic hormones are provided by Gade *et al.* (1997) and Coast *et al.* (2002). A briefer analysis of recent work on endocrine control of salt balance in insects has been written by Coast (2007), and Orchard (2006) has reviewed the orchestration of diuresis and other physiological events in the hemipteran *Rhodnius* by the amine serotonin. Much of what we know about the functioning of the insect hindgut is due to the extensive work of Phillips and co-workers (Phillips *et al.*, 1996). The dominant role of the vacuolar-type H^+-ATPase in driving solute transport across plasma membranes of insect epithelial cells has been reviewed frequently (e.g. Wieczorek *et al.*, 2000; Beyenbach and Wieczorek, 2006). The mechanism and regulation of secretory transport in Malpighian tubules of the yellow fever mosquito *Aedes aegypti* has been reviewed by Beyenbach (2003). The application of genetic, molecular biological and physiological tools has lead to tremendous advances in our understanding of the mechanisms and control of solute and water transport by the Malpighian tubules of the fruit fly *Drosophila melanogaster*. This area has been reviewed frequently, primarily by Julian Dow and Shireen Davies (Dow and Davies, 2003, 2006; Davies and Day, 2006).

2 Design principles underlying the insect excretory system

The insect excretory system is comprised of the Malpighian tubules and the gut, especially the hindgut (Fig. 1). Each Malpighian tubule consists of a single-layer of squamous epithelial cells which form a blind-ended tube. Tubules range from 2 to 70 mm in length, 2–250 in number and up to 100 μm in diameter (Phillips, 1981). Transport of ions (primarily K^+, Na^+ and Cl^-) and osmotically obliged water into the tubule lumen produces a near isosmotic secreted fluid. Fluid secretion is accompanied by passive diffusion of small solutes into the lumen, as well as selective secretion of solutes, including toxins. Some KCl and water may be reabsorbed downstream in a proximal segment of the Malpighian tubule or the anterior hindgut. The bulk of water, ion and metabolite reabsorption occurs downstream in the posterior hindgut, in particular the rectum, resulting in strongly hyperosmotic or hypoosmotic excreta. In many terrestrial species of insect, the hindgut can recover virtually all the water from the gut contents and fluid secreted into the gut by the Malpighian tubules. So effective are these water recovery systems that in the mealworm *Tenebrio* (Ramsay, 1971) and in the firebrat *Thermobia* (Noble-Nesbitt, 1970), for

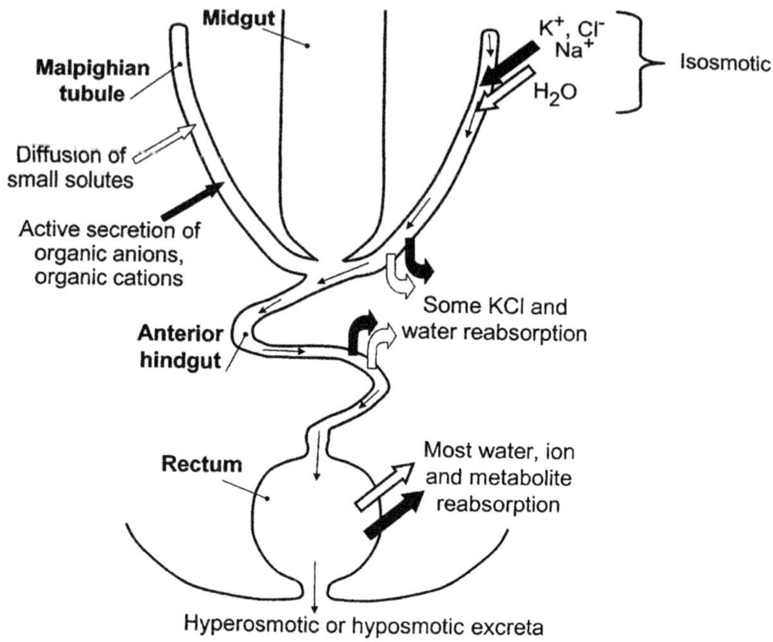

FIG. 1 Schematic diagram of the insect excretory system. Active transport is indicated by solid arrows and passive transport is indicated by open arrows. The flow of fluid secreted by the Malpighian tubules is indicated by the thin arrows. The figure is based on that of Phillips (1981, Fig. 1) with modifications.

example, the fecal pellets are powder dry. In addition, water vapour can be taken up from moist air drawn into the hindgut in both these species (O'Donnell and Machin, 1988). An essential feature of the insect hindgut for such extreme water reabsorption is the presence of a cuticular lining. The hindgut cells are thus protected by this cuticular lining from high concentrations of solutes, wastes and toxins in the lumen during water recovery (Phillips and Dockrill, 1968).

Adjustments of hemolymph fluid and solute levels are required in response to processes such as desiccation, feeding or osmotic influx across the external surfaces of aquatic species. However, several features of insect physiology and morphology mean that there has been less evolutionary pressure for rapid control of hemolymph composition by the excretory system. Firstly, respiratory gas exchange is accomplished by the ramifying tubes of the tracheal system whose termini extend into cells and tissues. Insects are not reliant, therefore, upon a blood-borne respiratory pigment whose functions would be compromised by changes in the volume and composition of the extracellular fluids. Secondly, sensitive tissues such as the eyes, the testis and the central nervous system are protected from the extracellular fluids by the actions of a series of epithelia, characterized by the presence of simple tight junctions. Such epithelia are not found in other invertebrates, but they play a crucial homeostatic role for the underlying tissues in insects. The regulatory properties of the perineurium, which envelopes the insect central nervous system, have been particularly well described (e.g. Treherne *et al.*, 1984). Thirdly, one of the pre-eminent adaptations of insects for terrestrial life, is the possession of a waxy cuticle. A consequence of the small size of insects is that they have a very large ratio of surface area to volume and they are thus faced with the potential for relatively rapid and large scale changes in the composition of their extracellular fluids. Insects are commonly found in osmotically and ionically stressful environments and life on land, in particular, poses the threat of desiccation. The waxy cuticle is thus a feature which limits water loss.

A key role of an excretory system is to eliminate automatically toxins not encountered before. Given the large number of plant-derived toxins which have evolved to minimize herbivory, this requirement is particularly important in insects. Arthur Ramsay was among the first to appreciate that the Malpighian tubule provides a means for removal from the hemolymph of all soluble substances of low molecular weight, and that in this respect it has analogies with both the glomerulus and the tubule of the vertebrate nephron. His classic paper (Ramsay, 1958) examined the diffusion of organic solutes such as sugars, urea and amino acids across the Malpighian tubule wall of the stick insect *Carausius morosus*.

The view that emerged from this and subsequent work is that the absolute permeability of the Malpighian tubule epithelium of insects is quite low when compared to the excretory epithelia of other invertebrates and the vertebrate kidney. This arrangement seems paradoxical (O'Donnell *et al.*, 1984; Skaer *et al.*, 1987), given the need to clear toxic compounds from the hemolymph. The solution to this paradox lies in the restriction of the *area* for passive permeation, namely the paracellular clefts, rather than the restriction of the permeability of these sites. The adjacent cells in insect epithelia such as the Malpighian tubule are separated by septate junctions rather than the occluding, or tight, junctions of transporting epithelia in vertebrates (Lane and Skaer, 1980). An intercellular space which is about 17 nm in width separates the large flattened cells of the Malpighian tubules of *Rhodnius*, for example. The ratio of the areas of the basal side of the cells (i.e. the transcellular route) and the intercellular cleft is 3000:1. However, when the 40-fold amplification of basal surface area due to membrane folding and convolution is accounted for, the ratio of transcellular to paracellular area increases to 120 000:1 (Skaer *et al.*, 1987). Compounds which are not transported actively diffuse through the intercellular clefts and are excreted; useful components also move slowly, and can be recovered from the excretory fluid by active transport.

Molecules cross the tubule epithelium by passive diffusion at rates which vary inversely with molecular mass (Maddrell and Gardiner, 1974;

Dalton and Windmill, 1981). Molecular mass also determines the route of transport for solutes which are not actively transported. Uncharged solutes whose molecular mass is less than that of sucrose cross into the lumen of the upper Malpighian tubule of *Rhodnius* by a cellular route (O'Donnell *et al.*, 1984). Surprisingly, mannitol, a relatively large molecule which has often been used as an extracellular space marker, readily permeates through the cells and enters the lumen about 10 times more rapidly than sucrose, even though this route entails crossing at least two cell membranes in series. Rapid permeation by the transcellular route in *Rhodnius* presumably reflects the large area of cell membrane available, as noted above. Uncharged solutes which have a molecular mass greater than that of sucrose as well as charged solutes are excluded from the tubule cells and can cross only via the intercellular junctions.

It is important to emphasize the relatively non-selective nature of the filtration process by any excretory system. The vertebrate glomerulus passes compounds as large as inulin, whose molecular weight is ~5000, and the coelomosac of the crustacean antennal gland is even more permeable, passing molecules as large as 50 000 Da. The pill millipede, *Glomeris marginata* is not an insect but it does have Malpighian tubules which rapidly secrete fluid very similar in composition to the bathing fluid (Farquharson, 1974). Even large solutes as inulin and dextrans of molecular weight up to 16 000 appear in the secreted fluid at concentrations only slightly below those of the bathing fluid (Farquharson, 1974).

In sharp contrast, the Malpighian tubules of insects secrete fluid which contains even such small molecules as sugars and amino acids at concentrations much below those in the hemolymph. For example, during diuresis the Malpighian tubules of the blood-feeder *Rhodnius prolixus* secrete fluids which contain glycine and inulin at concentrations only 1% of those in the bathing fluids. This low overall permeability of the tubule wall is a consequence of the limited area of the permeable sites rather than by restriction at the sites of permeation. As pointed out by Maddrell (1981) this arrangement means that toxic molecules, even those of moderate size, will still be removed passively, albeit slowly, from the insect's hemolymph. By contrast, the tubules can secrete fluid at extraordinary rates in the presence diuretic factors, as described below.

In sum, the excretory system of insects operates relatively slowly and the passive rate of movement of hemolymph solutes into a slowly secreted primary excretory fluid is restricted by a reduction in the area available for passive transfer. Maddrell (1981) has pointed out several advantages of this arrangement. Firstly, less energy is required both to produce the primary excretory fluid and to reabsorb useful substances from it. Secondly, relatively high concentrations of molecules such as amino acids, trehalose and lipids can be maintained in circulation. Flying insects use the disaccharide trehalose and/or diglyceride lipid in the hemolymph as fuels for flight, and these compounds may be present at concentrations as high as 100 mM (Weis-Fogh, 1967). Many species in the orders Lepidoptera, Hymenoptera and Coleoptera contain levels of amino acids in the hemolymph as high as 100–200 mM (Maddrell, 1971). Hormones that are released into circulation will also be removed slowly by the excretory system and the amounts which must be released may therefore be reduced.

During diuresis, insects can eliminate excess fluid at very high rates and yet lose only trace amounts of hemolymph solutes. Comparisons presented in Maddrell (1981) show that most vertebrates filter their extracellular fluid 10–20 times more rapidly than do most insects. Given that many hemolymph solutes of even quite small molecular size only appear in the filtered fluid at concentrations in the range 10–50% of those in the insect's hemolymph, reabsorption of such solutes need only occur in insects at about 1–2% of the rate required in many vertebrates, and there is thus a large saving in the energetic costs associated with reabsorption.

This energy saving may be a driving force behind the evolution of very rapid rates of fluid secretion by Malpighian tubules stimulated with diuretic factors When fluid is secreted at high rates, dilution of solutes in the tubule lumen is associated with reduced rates of diffusion back into the hemolymph. However, once the concentration of a solute in the lumen reaches a very low level, further increases in the rate of fluid secretion have virtually no more effect on solute entry. If excess fluid is very quickly removed, and contains only low levels of hemolymph solutes, no reabsorption of the solutes is needed. This may explain in part why many blood-sucking insects can excrete fluid at such extraordinarily high rates; after feeding the hemipteran *Rhodnius* can eliminate fluid equivalent in volume to its total hemolymph in as little as 30 min.

Passive diffusion of solutes from hemolymph to tubule lumen is backed up by active mechanisms that can rapidly transport a very wide range of substances that insects repeatedly need to reabsorb, such as glucose (Knowles, 1975a), or excrete, such as organic anions, urate, phosphate, magnesium, sulfate, sulfonates, acylamides, alkaloids and glycosides (Maddrell, 1977; O'Donnell *et al.*, 1983; Linton and O'Donnell, 2000; Torrie *et al.*, 2004). The importance of specific transcellular transporters for many classes of potential toxins has been emphasized in a DNA microarray study of Malpighian tubule-specific gene expression by Wang *et al.* (2004). There is high enrichment (9- to 29-fold) of 3 of the 5 organic anion transporter genes in the tubule-specific microarray, relative to whole flies. Similarly, the genes for 11 of 21 organic cation transporters are enriched 4–37-fold in the tubules relative to whole flies. The prime significance of this study is that it suggests a re-interpretation of the major functions of the *Drosophila* tubule. The epithelium is now viewed less as one primarily concerned with the secretion of inorganic salts and water and more as an epithelium richly endowed with solute transporters, including transporters for many potential toxins. An additional significance of the microarray study is that enrichment of genes for transporters for so many organic and inorganic solutes implies that a non-specific paracellular pathway plays only a minor role for many toxins (Wang *et al.*, 2004; Dow and Davies, 2006).

3 Secretion of physiological ions by insect Malpighian tubules

The hemipteran *R. prolixus* provides a useful model species for studying excretion of physiological ions and water that are present in excess of the insect's needs. *Rhodnius* ingests blood meals which may exceed the animal's unfed weight by a factor of 10 and during the subsequent diuresis it voids urine at a rate equal to its original body weight every 20–30 min. The prodigious rate of urine production means that each cell in the secretory segment of the Malpighian tubule secretes a volume of fluid equal to its own volume every 15 s (Maddrell, 1991) and cellular Cl^- content is exchanged every 5 s (O'Donnell *et al.*, 2003). These extraordinary rates of ion and water transport, coupled with the large size of the cells, ~100 μm in length, make the *Rhodnius* Malpighian tubule highly suitable for studies of epithelial transport.

The major transporters and pathways for movements of Na^+, K^+ and Cl^- across the upper (secretory) segment of the Malpighian tubule of the hemipteran *R. prolixus* are shown in Fig. 2. This model of fluid secretion is based on data from electrophysiological recordings using intracellular and ion-selective microelectrodes, the effects of ion substitution and transport inhibitors and also on immunofluorescent localization of transporters. Tubules of *Rhodnius* and the ant *Formica polyctena* are composed of a single cell type in the secretory segment. In the following sections, each of the inorganic ion transporters in the *Rhodnius* tubule as well as the paracellular pathway are discussed separately. Tubules of dipteran insects such as mosquitoes and fruit flies, as well as those of orthopterans such as the locusts *Locusta migratoria* and *Schistocerca gregaria*, contain more than one cell type in the secretory segment. The arrangements of ion transporters and ion permeation pathways for these tubules are discussed in Section 2.2. The control of the Malpighian tubules and hindgut by diuretic and antidiuretic hormones has been extensively reviewed by Coast *et al.* (2002), and is therefore considered only briefly in the sections below.

3.1 ION TRANSPORT ACROSS MALPIGHIAN TUBULES COMPOSED OF A SINGLE CELL TYPE

The blood upon which *Rhodnius* feeds has an osmolality of 320 mOsm kg^{-1}, approximately 14% hypoosmotic to the insects hemolymph osmolality of

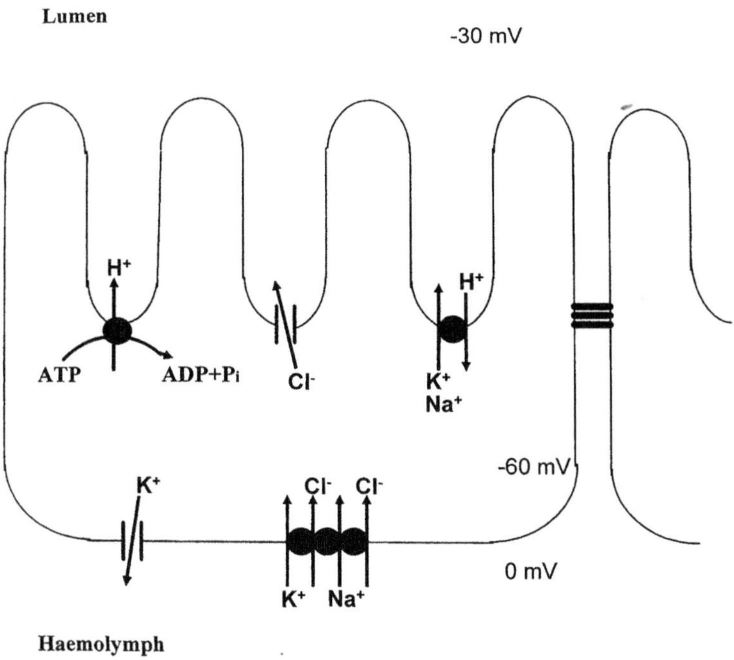

FIG. 2 Schematic diagram of the current model for transepithelial ion transport by cells in the upper (secretory) segment of the *Rhodnius* Malpighian tubule. Basolateral and transepithelial potentials are indicated (Ianowski *et al.*, 2004, Fig. 1). Reprinted with permission from the Company of Biologists.

370 mOsm kg^{-1} (Maddrell and Phillips, 1975b). Homeostasis therefore requires production of a hypoosmotic urine through a two-stage process involving secretion of Na^+, K^+, Cl^- and water across the upper Malpighian tubule followed by reabsorption of K^+ and Cl^- but not water downstream in the lower Malpighian tubule, as described in Section 4.1. The fluid voided from the insect during diuresis contains ~120 mM NaCl, matching the composition of the plasma fraction of the blood.

The fluid secreted by the upper segment of Malpighian tubules of *R. prolixus* during diuresis consists of approximately 100 mM NaCl and 80 mM KCl (Maddrell and Phillips, 1975b). Secretion of ions and osmotically obliged water by tubules of *Rhodnius* and other species is driven primarily by an apical vacuolar-type H^+-ATPase (Wieczorek *et al.*, 1989; Maddrell and O'Donnell, 1992). As a general rule, the H^+-ATPase plays a dominant role in energizing transport across insect epithelia, in contrast to the role of the Na^+/K^+-ATPase in vertebrate epithelia. This difference may reflect the co-evolution of insects with plants, which provide a K^+-rich and Na^+-poor food source. It is suggested that electrogenic transport of H^+ from the cell to the lumen energizes amiloride-sensitive exchange of cytoplasmic K^+ and/or Na^+ for luminal H^+ (Fig. 2). Entry of Na^+, K^+ and Cl^- through a basolateral Na^+–K^+–$2Cl^-$ cotransporter has been proposed for tubules of *R. prolixus* on the basis of the effects of bumetanide, Na^+-free saline and Cl^--free saline on fluid secretion and transepithelial potential (O'Donnell and Maddrell, 1984; Ianowski and O'Donnell, 2001). Na^+:K^+:$2Cl^-$ cotransport has also been implicated in basolateral entry of ions into Malpighian tubules of other species, including *Ae. aegypti* (Hegarty *et al.*, 1991), *F. polyctena* (Leyssens *et al.*, 1994), *Manduca sexta* (Audsley *et al.*, 1993; Reagan, 1995), *Teleogryllus oceanicus* (Xu and Marshall, 1999b) and *L. migratoria* (Al-Fifi *et al.*, 1998).

The Na^+/K^+-ATPase and K^+ channels, although present in the basolateral membrane of the upper tubule, play a minor role in net transtubular transport during diuresis in *Rhodnius*. In tubules of the ant *F. polyctena*, however, K^+ channels play a significant role in transepithelial transport when hemolymph K^+ levels are high, as described below in Section 3.1.5.

Measurements of transepithelial, basolateral and apical membrane potential have been used to identify the sequential activation of ion transporters when *Rhodnius* Malpighian tubules are stimulated with 5-hydroxyryptamine. A characteristic triphasic change in transepithelial electrical potential (TEP) has been reported when fluid secretion by isolated Malpighian tubules of *R. prolixus* is stimulated by serotonin (O'Donnell and Maddrell, 1984). From an initially negative value in unstimulated tubules (−2.5 mV, lumen-negative), TEP shifts to −33 mV in phase 1, +30 mV in phase 2 and −32 mV in phase 3 (Fig. 3). TEP was measured in the study of O'Donnell and Maddrell (1984) using the Ramsay technique, in which electrodes are positioned in bathing and secreted fluid droplets for tubules isolated under paraffin oil. The validity of this method of TEP measurement has been questioned on the grounds that, in tubules of small diameter, it may permit shunting of current from lumen to bath through the cells or through the thin layer of fluid adherent to the surface of that portion of the tubule in the oil (Aneshansley *et al.*, 1988; Isaacson and Nicolson, 1989). TEP measurements using the Ramsay technique were therefore compared with two other methods in which the voltage-sensing electrode is positioned in the tubule lumen by impalement of the Malpighian tubule with a microelectrode or by means of a cannula advanced into the tubule (Ianowski and O'Donnell, 2001). TEP measured during each phase of the stimulation with 5-hydroxytryptamine was similar irrespective of the measurement technique, validating the original measurements of O'Donnell and Maddrell (1984). The large diameter of the *Rhodnius* tubule (~100 μm) presumably minimizes the shunting of current described in the smaller diameter tubules of species such as *Ae. aegypti*.

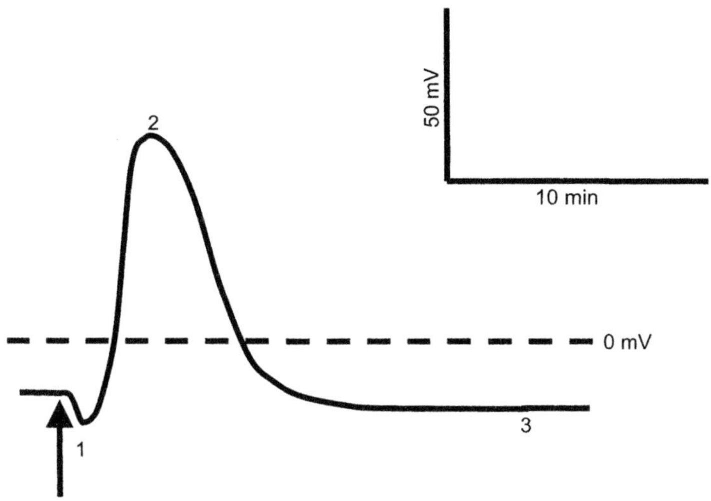

FIG. 3 Changes in transepithelial potential of the *Rhodnius* Malpighian tubule in response to stimulation with serotonin (5-HT). The current model suggests that the triphasic response corresponds to successive activation of (1) an apical Cl^- conductance, (2) the apical V-type H^+-ATPase and (3) the basolateral Na^+:K^+:$2Cl^-$ cotransporter. Based on data in Ianowski and O'Donnell (2001).

Ion substitution experiments and the effects of specific pharmacological reagents were used to support the proposal that the three phases of the response of TEP to serotonin correspond to sequential activation of an apical Cl^- channel, an apical V-type H^+-ATPase and a basolateral Na^+:K^+:$2Cl^-$ cotransporter (Fig. 3). Pre-incubation of tubules in Cl^--free saline abolishes phase 1, whereas pre-incubation in the presence of the H^+-ATPase inhibitor bafilomycin abolishes phase 2. Pre-incubation of the tubules in saline containing bumetanide, an inhibitor of Na^+:K^+:$2Cl^-$ cotransporters, abolishes phase 3 but does not alter phases 1 and 2. Negative going shifts in TEP during the first and third phases reflect enhanced movement of Cl^- across the apical membrane from cell to lumen. An increase in apical Cl^- conductance during phase 1 could account for this change in TEP. The transepithelial potential will thus reflect contributions of the V-type H^+-ATPase, tending to drive the lumen to more positive values, and compensating movement of Cl^- from cell to lumen, tending to drive the lumen to more negative values. Once the apical permeability to Cl^- has increased in phase 1, the apical membrane potential will be influenced by changes in intracellular Cl^- activity following enhanced entry of Cl^- into the cell by activation of a basolateral Na^+:K^+:$2Cl^-$ cotransporter in phase 3.

3.1.1 *The apical vacuolar H^+-ATPase*

The structure, function and regulation of insect and other V-ATPases has been reviewed extensively in recent years, primarily by Wieczorek, Harvey and co-workers (Harvey and Wieczorek, 1997; Wieczorek *et al.*, 2000). Beyenbach and Wieczorek (2006) provide an historical overview in a recent review. V-ATPases are multisubunit, heteromeric proteins which consist of two structural domains, a peripheral, catalytic V_1 domain ($\sim$ 500 kDa) and a membrane-spanning V_o domain (100–250 kDa). Regulation of V-ATPase activity may be accomplished by multiple mechanisms, including disassembly/reassembly of V_1 and V_o domains, oxidation of SH groups, complexation with activator or inhibitor proteins or small signalling molecules, and sorting of the holoenzyme or its subunits to target membranes (Merzendorfer *et al.*, 1997). The first gene knockout of a V-ATPase in any animal was accomplished in *Drosophila* (Davies *et al.*, 1996). This study reported that epithelial dysfunction may be detectable at an early stage in those tissues where V-ATPases play a plasma membrane role and that the genes are essential for normal embryonic development (*ibid.*) An experimental and *in silico* survey of the V-ATPase gene family in *Drosophila* has revealed that 13 of the more than 32 genes are specialized for epithelial roles. When these genes are disrupted, they display a clear tubule phenotype that is conserved across mutant alleles for almost all the V-ATPase subunits; the transparent tubules result from a loss of birefringence in the tubule lumen and is a consequence of a failure to excrete uric acid (Allan *et al.*, 2005).

V-ATPases transduce the energy from ATP hydrolysis into a proton current across a membrane. The resultant electrochemical gradient $\Delta\mu_H$ is termed the proton motive force (PMF) and it is used to drive the movements of many other ions through carriers and channels. V-ATPases are inherently electrogenic because their activity pumps protons without an associated anion. In the absence of other ion carriers, the V-ATPase will generate a large voltage across a membrane and produce a modest change in H^+ concentration. In the presence of anion channels (e.g. Cl^-) a large acid flux is generated. By contrast, an alkalizing flux is generated in the presence of a cation: nH^+ antiporter or symporter and appropriate counterions (anions).

In contrast to P-type-ATPases, such as the Na^+/K^+-ATPase and the mammalian gastric H^+-ATPase, V-ATPases do not form a phosphorylated intermediate and are relatively insensitive to vanadate. They are sensitive to the potent and specific inhibitor bafilomycin A_1, a macrolide antibiotic isolated from the fermentation of the Gram-positive actinobacteria *Streptomyces* spp. (Drose and Altendorf, 1997). Fluid secretion by the tubules of *Rhodnius* is inhibited 49% by 1 μM bafilomycin (Gutierrez *et al.*, 2004), for example, and secretion by

F. polyctena (Leyssens *et al.*, 1994), the cricket *Te. oceanicus* (Xu and Marshall, 1999b) and the hemipteran *R. prolixus* (Ianowski *et al.*, 2002). Measurements with ion-selective microelectrodes for Na^+, K^+ and Cl^- show that Na^+:K^+:$2Cl^-$ cotransport is thermodynamically feasible in both unstimulated and 5HT-stimulated *Rhodnius* tubules (Ianowski *et al.*, 2002). Chloride activity in the cell is higher than expected on the basis of passive distribution of Cl^- across the basolateral membrane in both unstimulated and 5HT-stimulated cells (Fig. 5). Cl^- must therefore be actively transported across the basolateral membrane into the cell. K^+:Cl^- involvement in net transepithelial ion transport can be ruled out on the basis of these measurements, because the net electrochemical potential for such a transporter is positive, indicating that it would mediate net movement of K^+ and Cl^- from inside the cell to the bath, opposite to the direction of transepithelial transport during fluid secretion.

Operation of the Na^+:K^+:$2Cl^-$ cotransporter at a low rate in unstimulated tubules was suggested by Maddrell and Overton (1988) and is consistent with measured net electrochemical potentials. In unstimulated tubules, K^+ may also enter cells from the bath both through an ouabain-sensitive Na^+/K^+-ATPase (Maddrell and Overton, 1988). Addition of ouabain to unstimulated tubules reduces one path for K^+ entry and blocks the transport of Na^+ from cell to bath, with a resulting increase in Na^+ transport from cell to lumen (Maddrell and Overton, 1988).

Transepithelial ion transport increases nearly 1000-fold in response to serotonin. Stimulation increases both the apical ion transporters as well as

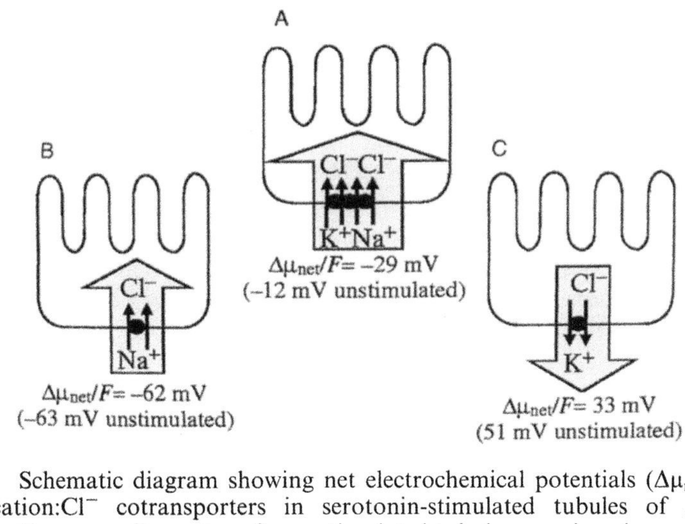

FIG. 5 Schematic diagram showing net electrochemical potentials ($\Delta\mu_{net}/F$) for three cation:Cl^- cotransporters in serotonin-stimulated tubules of *Rhodnius prolixus*. Corresponding values for unstimulated tubules are given in parentheses. (A) Na^+: K^+:$2Cl^-$, (B) Na^+:Cl^- and (C) K^+:Cl^- (Ianowski *et al.*, 2002, Fig. 2). Reprinted with permission from the Company of Biologists.

the basolateral Na^+:K^+:$2Cl^-$ cotransporter (Maddrell and Overton, 1988; Maddrell, 1991). Under these conditions ion flux through the basolateral cotransporter is so much greater than that through the Na^+/K^+-ATPase that fluid secretion in stimulated tubules is insensitive to ouabain.

An intriguing property of the basolateral cation:Cl^- cotransporter in *Rhodnius* Malpighian tubules is the ability for K^+ to be replaced by Na^+. The study of Ianowski *et al.* (2004) is the first description of competitive inhibition of K^+ transport by Na^+ in a bumetanide-sensitive mechanism. The likely function of K^+ replacement by Na^+ is to permit homeostatic regulation of hemolymph K^+ levels even when tubules are secreting fluid at near maximal rates. Although tubules bathed in saline containing 6 or 10 mM K^+ and 137 mM Na^+ secrete fluid at equal rates, the fluxes of Na^+ and K^+ differ greatly. For tubules in saline containing 6 mM K^+, Na^+ flux is > threefold that of K^+, whereas the fluxes of Na^+ and K^+ are equal in saline containing 10 mM K^+. Total cation (Na^+ and K^+) flux is the same in the two salines. Moreover, addition of bumetanide reduces fluid secretion rate to the same extent in both salines. These findings are consistent with replacement of K^+ by Na^+ in a bumetanide-sensitive cotransport mechanism during serotonin-stimulated fluid secretion. In the extreme case of complete removal of K^+ from the bathing saline, Malpighian tubules secrete fluid at 45% of the rate in the control saline and remain sensitive to bumetanide. However, transport is unaltered by the drug hydrochlorothiazide, an inhibitor of Na^+:Cl^- cotransporters in vertebrate tissues. Dose–response curves relating percentage inhibition of fluid secretion to bumetanide concentration are identical for tubules bathed in control and K^+-free saline, suggesting that fluid secretion involves the same bumetanide-sensitive cotransport system in the presence or absence of K^+. Malpighian tubules bathed in K^+-free saline also show the characteristic triphasic change of apical membrane potential in response to stimulation with serotonin. Addition of bumetanide to such tubules produces a lumen-positive shift in apical membrane potential, as described above for tubules in control saline. Taken together, measurements of fluid secretion rates, cation fluxes and apical membrane potentials indicate that Na^+ can replace K^+ in a single bumetanide-sensitive cotransport mechanism in the basolateral membrane.

Kinetic analysis also supports the idea that Na^+ competes with K^+ for transport by a basolateral bumetanide-sensitive cation/Cl^- cotransporter during fluid secretion. Increasing Na^+ concentration in the bath increases the Michaelis–Menten constant (K_t) for K^+ transport while the maximum transepithelial K^+ flux (J_{max}) remains constant, consistent with competitive inhibition of K^+ transport by Na^+. This is in stark contrast to the common finding that increasing bathing saline Na^+ concentration produces an increase in K^+ flux through bumetanide-sensitive $Na^+/K^+/2Cl^-$ cotransporters in vertebrate epithelia (for reviews, see Russell, 2000).

During diuresis, *Rhodnius* produces copious amounts of NaCl-rich urine, effectively eliminating the nutrient-poor plasma fraction of the blood meal and retaining the nutrient-rich blood cells. Studies by Maddrell *et al.* (1993) showed that the upper tubule responds to reductions in hemolymph K^+ concentration by reducing the K^+ concentration in the secreted fluid. This reduction enhances reabsorption of K^+ by the lower tubule, thereby contributing to homeostatic regulation of hemolymph K^+. Competition between Na^+ and K^+ for transport by a single bumetanide-sensitive cotransporter in the upper tubule provides a mechanism for reducing secreted fluid K^+ concentration, thereby reducing the need for K^+ reabsorption by the lower Malpighian tubule. Importantly, the increase in Na^+ flux when saline K^+ concentration is reduced not only minimizes the loss of K^+ but at the same time permits the rate of urine production to remain very high.

The contribution of K^+ channels to transport of K^+ from bath to lumen during fluid secretion by *R. prolixus* tubules can be ruled out on the basis of two types of evidence. First, treatment with Ba^{2+} does not inhibit fluid secretion, suggesting that K^+ channels are not a significant component of transepithelial K^+ transport at physiological concentrations of extracellular K^+. Second, basolateral membrane potential depolarises in response to addition of Ba^{2+} (Ianowski *et al.*, 2002). Depolarization is consistent with blockage of K^+ leakage into the bath, since the basolateral membrane potential appears to be determined primarily by the K^+ conductance (i.e. basolateral membrane potential is very similar to the Nernst equilibrium potential for potassium, E_K). The depolarisation of the basolateral membrane potential after addition of Ba^{2+} is consistent with a small but positive value for $\Delta\mu_K/F$ that favours movement of K^+ from cell to bath through ion channels.

3.1.4 *Transport of Cl^- across the apical membrane*

Intracellular chloride activity in *Rhodnius* tubules declines when the apical membrane potential increases (i.e. as the transepithelial potential becomes more lumen-positive) in Na^+- or K^+-free saline or in the presence of bumetanide (Ianowski *et al.*, 2002). This finding is consistent with the presence of apical Cl^- channels which would mediate transport of Cl^- from cell to lumen when entry of Cl^- into the cell across the basolateral membrane is blocked. In Na^+-replete saline, intracellular Cl^- activity is near equilibrium across the apical membrane both in unstimulated and in serotonin-stimulated tubules (Fig. 4A, B), as would be expected if Cl^- channels mediate vectorial ion transport across the apical membrane. Cl^- would be expected to be very close to electrochemical equilibrium across the apical membrane if Cl^- channels account for most of the apical membrane conductance. Although 4,4′-diisothiocyano-2,2′- disulfonic acid stilbene (DIDS) inhibits fluid secretion by *Rhodnius* tubules, it has no effect on transepithelial potential or intracellular pH, as would be expected if DIDS blocked apical Cl^- channels or Cl^-/HCO_3^- exchangers, respectively. This suggests that Cl^- channels, if present, are DIDS-insensitive and that the effects of fluid DIDS on fluid secretion may reflect indirect effects of DIDS, such as inhibition of mitochondrial functioning. It is worth pointing out that the possibility of a DIDS-sensitive channel mediating Cl^- entry into the cells across the basolateral membrane can be ruled out on the basis of the measured electrochemical gradients described above. Chloride activity is above electrochemical equilibrium, and the opening of basolateral Cl^- channel would result in an efflux rather than an uptake of Cl^-.

The contribution of other pathways for Cl^- movement, such as apical K^+:Cl^- cotransport and/or paracellular Cl^- movement, can be ruled out because they are thermodynamically unfeasible. Although the transepithelial potential across tubules of most insects is lumen-positive, the TEP in fully stimulated *Rhodnius* tubules is lumen-negative and a passive paracellular pathway for Cl^- movement is thus precluded. The predicted Nernst equilibrium activity of Cl^- in the lumen for a bathing saline Cl^- activity of 93 mM (Ianowski *et al.*, 2002) and a lumen negative transepithelial potential of –28 mV is 26 mM, which is ~3.5-fold less than the measured activity of 109 mM (Ianowski and O'Donnell, 2006). The calculated electrochemical gradients also show that if a K^+:Cl^- cotransporter were active in the apical membrane of *Rhodnius* Malpighian tubules, it would mediate net movement of these ions from the lumen into the cell, in the opposite direction to that required for fluid secretion. A proposal of K^+:Cl^- cotransport from cell to lumen was based on the inhibition of fluid secretion by dihydroindenyl)oxy]alkanoic acid (Gutierrez *et al.*, 2004). However this drug is also known to cause mitochondrial damage (Pond *et al.*, 2004). In addition, the experiments involved perfusion with fluid containing 8 mM K^+ (Gutierrez *et al.*, 2004) The potassium activity of this solution is ~ 6 mM, more than 8 times less than the physiological level of 49 mM in the lumen of 5HT-stimulated tubules.

3.1.5 *K^+ channels*

Although K^+ channels are not implicated in net transepithelial K^+ and fluid secretion by *Rhodnius* Malpighian tubules, they do appear to be important in tubules of the ant *F. polyctena* under some conditions. Multiple mechanisms of K^+ transport by tubules of this species may be necessary because hemolymph K^+ levels are quite variable. A study by

Van Kerkhove *et al.* (1989) showed that the hemolymph K^+ concentration in ants of one nest ranged from 6 to 20 mM, whereas in the hemolymph of ants from another nest the potassium concentration ranged from 12 to 70 mM. The Malpighian tubules may therefore be exposed to a wide range of K^+ concentrations *in vivo*, and this may explain the findings that a cascade of different mechanisms of K^+ uptake across the basolateral membrane operate as bathing saline K^+ concentration increases. The model of Leyssens *et al.* (1994) proposes that a Na^+:K^+:$2Cl^-$ cotransporter which is sensitive to 10^{-5} M bumetanide is the primary mechanism of uptake at the lowest bathing saline K^+ concentrations of 5–10 mM. At the intermediate concentration of 51 mM K^+, a K^+:Cl^- cotransporter sensitive to high (10^{-4} M) concentrations of bumetanide becomes active. At very high bathing saline concentrations (113 mM K^+) K^+ enters through basolateral Ba^{2+}-sensitive K^+ channels (Leyssens *et al.*, 1994).

A role for basolateral K^+ channels has also been proposed for the main segment of the tubules of the black field cricket *Te. oceanicus* (Xu and Marshall, 1999b). In this species as well as in the ant *F. polyctena*, hemolymph K^+ and Na^+ probably vary as a result of omnivory. Under some conditions, K^+ may enter the Malpighian tubule cells through Ba^{2+}-sensitive K^+ channels. At lower hemolymph K^+ concentrations, bumetanide inhibits fluid secretion by 80% and inhibits Na^+, K^+ and Cl^- secretion by 60%, 70% and 50%, respectively. This result suggests that a large proportion of net transepithelial ion transport is mediated by cotransport of these three ions (Xu and Marshall, 1999b).

3.2 ION TRANSPORT ACROSS MALPIGHIAN TUBULES COMPOSED OF PRINCIPAL CELLS AND STELLATE CELLS

The Malpighian tubules of dipterans such as the fruit fly *D. melanogaster* and the mosquito *Ae. aegypti* contain two cell types in the fluid secreting segment (Fig. 6). The structure and functioning of the Malpighian tubules of these species has been reviewed by Dow and Davies (2003) and by Beyenbach (2003). There is general agreement that cations and possibly some of the chloride moves through the larger principal cells which make up 80% of the cells. For the tubule secreting at the basal rate chloride is transported through the smaller stellate cells which comprise 20% of the cells in the secretory segment. When *D. melanogaster* tubules are stimulated with the amine tyramine or the peptides Drosokinin (Terhzaz *et al.*, 1999) or leucokinin, a pathway through chloride channels has been proposed. An alternative proposal based primarily on the analysis of the *Ae. aegypti* Malpighian tubule indicates a paracellular pathway that is stimulated by leucokinin. Roles for potassium channels, Na^+:K^+:$2Cl^-$ cotransporters and the Na^+/K^+-ATPase have been proposed for entry of

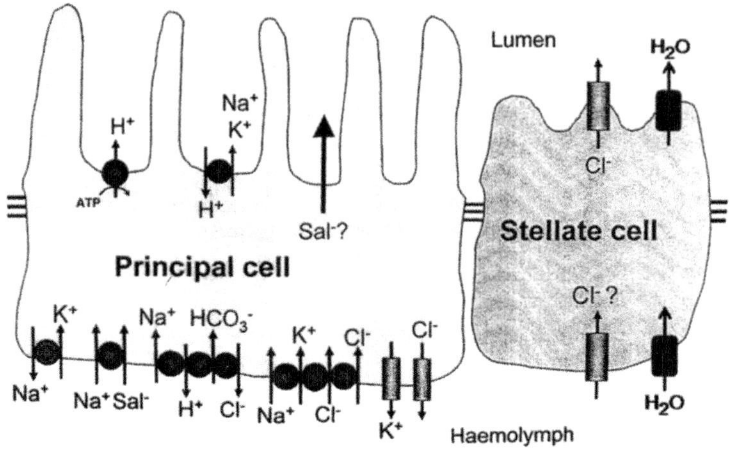

FIG. 6 Schematic diagram of ion transporters in the cells of the Malpighian tubules of *Drosophila melanogaster* (Ianowski and O'Donnell, 2004; O'Donnell *et al.*, 1996, 2003; Sciortino *et al.*, 2001). Studies by Kaufmann *et al.* (2005) show that water channels (DRIP) are expressed exclusively in the stellate cells. The uptake of salicylate across the basolateral membrane is mediated by a α-cyano-4-hydroxycinnamic acid-sensitive Na^+:salicylate cotransport system (Na^+:Sal^-). This transport system utilizes the inwardly Na^+ gradient created by the Na^+/K^+-ATPase to move salicylate into the cell as a secondary active transport process. The mechanism of movement of salicylate from cell to lumen (Sal^-?) is unknown (Ruiz-Sanchez and O'Donnell, 2006, Fig. 7). Reprinted with permission of Elsevier Science.

ions across the basolateral membrane of the principal cells. Ion movement from cell to lumen appears to involve much the same mechanisms as proposed for tubules of *Rhodnius* and *Formica*. An apical V-type H^+-ATPase drives protons from cell to lumen, and protons are then recycled back into the cell in exchange for potassium or sodium. The V-type proton pump has been extensively characterized in other insect epithelia such as the midgut of *Ma. sexta* (Beyenbach and Wieczorek, 2006) and the molecular characterization of the *D. melanogaster* Malpighian tubule V-type H^+-ATPase has been reviewed by Dow (1999) and Dow *et al.* (1997). The study of Allan *et al.* (2005) shows that the plasma membrane V-ATPase holoenzyme is a single isozyme and is composed of polypeptides derived from expression of 14 genes, not just in the Malpighian tubule but in all the major transporting epithelia (Allan *et al.*, 2005).

3.2.1 *The apical and basolateral cation: proton exchangers*

The nature of the putative amiloride-sensitive putative Na^+/H^+ and K^+/H^+ exchangers has been the subject of much recent attention. The sodium proton exchanger (NHE) gene family has been characterized in the *Drosophila* Malpighian tubules by Giannakou and Dow (2001). The order of inhibition of fluid secretion by amiloride derivatives (ethylisopropylamiloride ≫ 2,4-dichlorobenzamil > 5-*N*,*N*-dimethylamiloride > amiloride ≅ benzamil) is consistent with the presence of NHE rather than epithelial Na^+ channels (ENaC) in Malpighian tubules of *D. melanogaster* and *Ae. aegypti* (Petzel, 2000). The presence of Na^+/H^+ exchange is also consistent with RT-PCR data showing that NHE genes are expressed in Malpighian tubules of *Drosophila* whereas there is no evidence for expression of ENaC genes (Giannakou and Dow, 2001). The latter finding is consistent with the finding that amiloride inhibits transepithelial sodium secretion in *Ae. aegypti* tubules but does not alter membrane resistance, as would be expected for channel inhibition. Evidence presented by Petzel (2000) is most consistent with a basolateral location of NHE and a role in pH regulation. Malpighian tubules respond rapidly to changes in bathing saline pH and intracellular pH acidifies when sodium is removed from the bathing saline, consistent with the basolateral location of NHE. If NHE is located on the apical membrane and its function is to pump sodium into the lumen in exchange for protons, then exposure to sodium free bathing saline would be predicted to result in an alkalinization of the cell if the apical proton-ATPase continued to pump protons from cell to lumen.

The sodium/proton exchanger isoform 3 (NHE3) is localized to the basolateral membrane in tubules of *Ae. aegypti* and co-localizes with the Na^+/K^+-ATPase. Na^+ uptake in transgenic cells expressing this isoform is insensitive to amiloride (Pullikuth *et al.*, 2003, 2006). The authors propose that *Ae*NHE3 is not the exchanger operating in concert with the apical V-ATPase. However, *Ae*NHE3 may, by virtue of its basolateral localization, play a role in regulating intracellular pH that indirectly impacts V-ATPase and apical exchanger functions.

Kang'ethe *et al.* (2007) suggest that the apical Na^+/H^+ exchanger which transports cations into the lumen, driven secondarily by the proton gradient created by the V-type H^+-ATPase, is sodium/proton exchanger isoform 8 (NHE8). Immunolocalization studies show *Ae*NHE8 expression in the apical membranes of Malpighian tubules, gastric caecae, and rectum. Heterologous expression of *Ae*NHE8 in NHE-deficient fibroblast cells results in an amiloride-sensitive sodium uptake. When membranes from yeast cells expressing the protein are reconstituted into lipid proteoliposomes, fluorometric assays reveal saturable exchange of Na^+ and K^+ for H^+. *Ae*NHE8 may be coupled to the inward H^+ gradient across the Malpighian tubules and may play a role both in the extrusion of excess sodium and potassium and in maintenance of intracellular pH (Kang'ethe *et al.*, 2007).

An alternative explanation is that an electrophoretic antiporter ($Na^+/2H^+$) in the apical membrane of tubules and gut is driven by the electrical potential difference across the membrane created by the V-type H^+-ATPase (Rheault *et al.*, 2007). This would contrast the role of NHEs in catalyzing electroneutral Na^+/H^+ exchange using the Na^+ chemical gradient generated by the Na^+/K^+ P-ATPase. A cDNA encoding an electrophoretic antiporter has been cloned from the alimentary canal of larval African malaria mosquito, *Anopheles gambiae*, and is termed *Ag*NHA1. Immunolabeling of whole mounts and longitudinal sections of isolated alimentary canal show that *Ag*NHA1 is expressed in proximal Malpighian tubules and also in the cardia, gastric caeca, anterior midgut, posterior midgut and rectum, as well as in the suboesophageal and abdominal ganglia (Rheault *et al.*, 2007). *Ag*NHA1 is a likely candidate for the mosquito homolog of the caterpillar $K^+/2H^+$ antiporter identified by Wieczorek *et al.* (1991).

3.2.2 *Ancillary role of the Na^+/K^+-ATPase*

The role of the Na^+/K^+-ATPase in the basolateral membrane has long been controversial. In *Drosophila*, the Malpighian (renal) tubule contains large amounts of Na^+/K^+-ATPase that is known biochemically to be highly sensitive to ouabain, whereas the effects of the drug on the intact tissue appear to be variable and depend whether the tubules are secreting fluid at a basal rate or have been stimulated to secrete at higher rates by diuretic factors or cyclic AMP (cAMP). Ouabain, a specific inhibitor of the Na^+/K^+-ATPase, at concentrations of 10 μM depolarizes the electrical potential across the basolateral membrane (V_{bl}) and also causes a slight increase in the rate of fluid secretion by unstimulated tubules and an increase in the concentration of Na^+ in the secreted fluid (Fig. 7). The rapid depolarization of V_{bl} by ouabain is consistent both with inhibition of the contribution of an electrogenic pump ($3Na^+:2K^+$) and with a small decline in intracellular K^+ levels. An increase in the level of Na^+ within the cell in the presence of ouabain, in conjunction with the high Na^+ affinity of the apical cation/H^+ exchanger (Maddrell and O'Donnell, 1993), means that more Na^+ is transported into the lumen of the tubule, resulting in a higher Na^+ concentration in the secreted fluid and a slightly higher secretion rate (Fig. 7B, C). Similar effects were observed when ouabain was applied to the unstimulated Malpighian tubules of *R. prolixus* (Maddrell and Overton, 1988). In contrast, fluid secretion by unstimulated Malpighian tubules of *Ae. aegypti* (Hegarty *et al.*, 1991) and *L. migratoria* (Anstee *et al.*, 1979) is inhibited by 0.1–1 mM ouabain, as is secretion by cAMP-stimulated tubules of *Drosophila* (Linton and O'Donnell, 1999).

The Na^+/K^+-ATPase provides a minor route of K^+ entry into the tubule cells of *Drosophila* given that there is only a slight increase in the secretion rate and the $[Na^+]:[K^+]$ ratio of the secreted fluid when ouabain is applied. These data suggest that the functions of the Na^+/K^+-ATPase

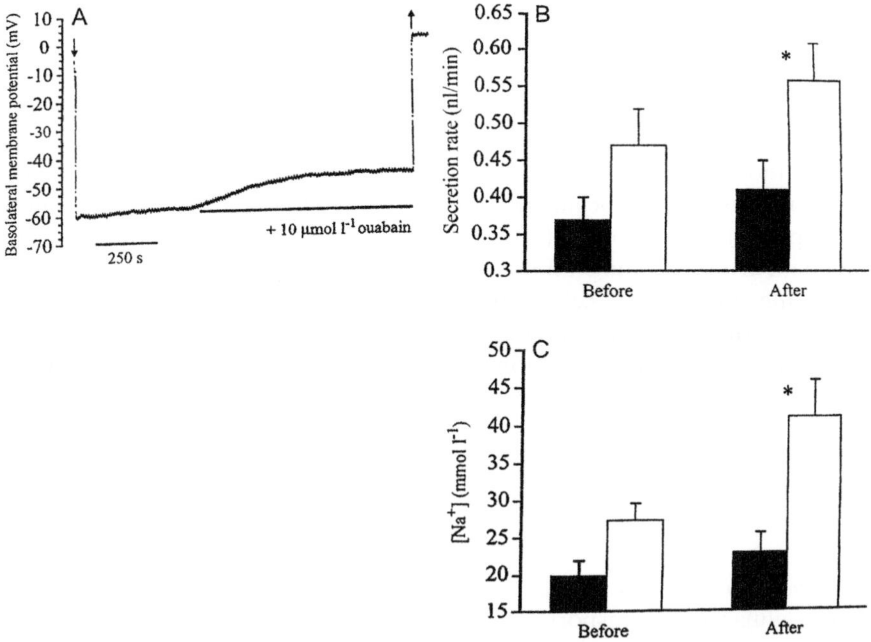

FIG. 7 Effects of 10 μM ouabain on (A) basolateral membrane potential (V_{bl}, mV), (B) secretion rate (nl min^{-1}) and (C) Na^+ concentration (mmol l^{-1}) of the fluid secreted by unstimulated Malpighian tubules in standard bathing medium (SBM). The sample recording in (A) shows the change in V_{bl} in response to the addition of ouabain for the period indicated by the horizontal bar. Downward and upward arrows indicate cell impalement and withdrawal of the microelectrode into the bathing saline, respectively. In (B) and (C) the values are means + S.E.M. ($n = 9$–16) before and 30 min after the addition of either ouabain (experimental group, open columns) or ethanol (0.1%; control group, filled columns). Asterisks indicate significant differences between the experimental and control groups ($P<0.05$). Redrawn from Linton and O'Donnell (1999, Figs. 4 and 5). Reprinted with permission from the Company of Biologists.

are to maintain the potential across the basolateral membrane by maintaining differential ion concentrations. Moreover, transport of Na^+ from the cell to the bath by the Na^+/K^+-ATPase may permit high levels of solutes to be accumulated by Na^+-coupled entry mechanisms with little loss of Na^+ by secretion into the lumen. Maintenance of low intracellular Na^+ concentrations by the Na^+/K^+-ATPase may be necessary in part because of the high affinity of apical cation/H^+ exchangers for Na^+ (Maddrell and O'Donnell, 1993). The concentration of ouabain required to inhibit fluid secretion by isolated Malpighian tubules is typically much greater than that required to inhibit ATP hydrolysis in subcellular fractions prepared from the same tubules. It appears that transport of ouabain into the lumen may lower the concentration of ouabain near the active sites in intact cells and that this effect can be minimized through the addition of competing substrates such as taurocholate for the ouabain transporter (Torrie *et al.*, 2004). However, it must be noted that measurements of the concentrations of organic anions and organic cations near the cell surface typically indicate concentrations only a few percent below those in the bulk medium (e.g. O'Donnell and Rheault, 2005). One possibility is that co-localization of the ouabain transporter organic anion transporting polypeptide (OATP) with the Na^+/K^+-ATPase prevents ouabain from reaching inhibitory concentrations within the basolateral infoldings of principal cells, although direct evidence for such an effect is lacking. An alternative explanation is that the competing substrates may indirectly inhibit fluid secretion, as occurs in response to high concentrations (> 0.5 mM) of organic anions such as salicylate (O'Donnell and Rheault, 2005). For example, 0.06 mM taurocholate inhibits fluid secretion by *Drosophila* tubules by ~25%, whereas in combination with 1 mM ouabain the effect is increased to 50% (Torrie *et al.*, 2004).

An additional complication is the finding of high levels of Na^+/K^+-ATPase in the stellate cells of the Malpighian tubules of the mosquito *Ae. aegypti* (Patrick *et al.*, 2006). The role of the stellate cells in transcellular Cl^- transport is discussed below in Section 3.2.4, but it is uncertain as to whether the cells are also involved in transcellular cation transport, or if the Na^+/K^+-ATPase plays a role in energizing ion transport for homeostatic functions such as pH regulation or cell volume regulation.

3.2.3 *Na^+:K^+:2Cl^- cotransporter*

Fluid secretion by the Malpighian tubules of *D. melanogaster* is inhibited approximately 25% by bumetanide (Linton and O'Donnell, 1999). When bumetanide alone is added to the saline bathing *Drosophila* tubules, K^+ flux decreases but Na^+ flux is unchanged. Under these conditions, however, much of the Na^+ that enters the cells is recycled back to the bathing saline by the Na^+/K^+-ATPase. Blockade of the Na^+/K^+-ATPase with ouabain increases secreted fluid Na^+ concentration from 23 to 45 mM, and subsequent addition of bumetanide reduces transepithelial fluxes of both Na^+ and K^+ by nearly equal amounts, consistent with inhibition of an Na^+:K^+:2Cl^- cotransporter (Ianowski and O'Donnell, 2004). Bumetanide also inhibits Na^+, K^+ and Cl^- secretion by *Aedes* tubules which have been stimulated with cAMP, consistent with the presence of a cAMP-stimulated Na^+:K^+:2Cl^- cotransporter (Hegarty *et al.*, 1991).

In tubules of both species, the Na^+ gradient across the basolateral membrane provides the driving force for secondary active Cl^- transport through a bumetanide-sensitive Na^+:K^+:2Cl^- cotransporter. The Na^+ gradient is maintained by the actions of two other transporters; the basolateral Na^+/K^+-ATPase and the apical Na^+/H^+ exchanger, which in turn is driven by the proton gradients established by the V-type H^+-ATPase. The Na^+/K^+-ATPase is thus poised to play a pivotal role in setting the Na^+/K^+ ratio of the secreted fluid. In particular, in unstimulated Malpighian tubules the basolateral Na^+/K^+-ATPase may return much of the Na^+ that enters the cell through Na^+-linked exchangers or cotransporters back to the hemolymph. The result will be high levels of K^+ in the secreted fluid. When the $Na^+/K^+/2Cl^-$ cotransporter is stimulated along with the apical vacuolar H^+-ATPase in response to increases in cellular levels of cAMP, there is an increase in the entry of both K^+ and Na^+ and in fluid secretion rate. As a result, the Na^+ content of the secreted fluid increases. This is particularly the case if the cotransporter is able to accept alternative stoichiometries. As noted above in Section 3.1.3, there is evidence that the alkali cation chloride cotransport or in the *R. prolixus* Malpighian tubule is able to accept Na^+ in place of K^+. As a result, the Na^+ content of the secreted fluid can increase above that of K^+. In addition, some of the K^+ which enters through the Na^+/ K^+-ATPase or through the alkali cation:chloride cotransporter may leak into the bathing saline through K^+ selective channels in the basolateral membrane.

3.2.4 *Na^+ conductance in the basolateral membrane of Aedes tubules*

In Malpighian tubules of *Aedes*, the natriuretic peptide or its second messenger cAMP activates both basolateral Na^+:K^+:$2Cl^-$ cotransport (Hegarty *et al.*, 1991) as well as a basolateral Na^+ conductance (Sawyer and Beyenbach, 1985). The depolarization of the basolateral membrane voltage together and the concomitant decline in transepithelial resistance and the fractional resistance of the basolateral membrane are consistent with an increase in the Na^+ conductance of the basolateral membrane. Enhanced entry of Na^+ into the principal cell presumably increases cytoplasmic [Na^+], thereby increasing its availability for extrusion across the apical membrane. The net effect is selective stimulation of transepithelial NaCl and water secretion.

3.2.5 *Cl^- channels and paracellular pathways*

Patch clamp studies have revealed two types of Cl^- channel in excised, inside-out apical membrane patches of the stellate cells of *Ae. aegypti*. Type I channels have a conductance of 24 pS, an open probability of 0.82 and a mean open time of 867 ms (O'Connor and Beyenbach, 2001). Type II Cl^- channels have a conductance of 8 pS, an open probability of 0.07 and a mean open time of 7.5 ms. The high density and halide selectivity sequence of the type I Cl^- channel is consistent with a role in transepithelial Cl^-

secretion under control conditions. The permeability of the tubule wall and type I channels share Eisenman halide selectivity sequence I (in order of decreasing permeability): $I^- > Br^- > Cl^- > F^-$ >isethionate. Type I channels, therefore, appear to be part of the transepithelial pathway taken by Cl^- under control conditions. In contrast, when stimulated by leucokinin, the tubule wall exhibits Eisenman selectivity sequence III: $Br^- > Cl^- > I^- > F^-$ > isethionate (Yu and Beyenbach, 2001), suggesting that type I channels are not part of the leucokinin-stimulated pathway. Excised patches show no effects of removal and/or chelation of Ca^{2+} upon type I currents, as might be expected if type I channels respond to the elevation of intracellular Ca^{2+} evoked by leucokinin. However, this effect may be due to loss of essential cell components (e.g. protein kinases) required to complete signal transduction in response to leucokinin. An essential future experiment will be to determine whether current through stellate cell ion channels increases in response to leucokinin when current is recorded using the cell-attached configuration.

DIDS-sensitive 'maxi' Cl^- channels with a conductance of 256 pS are found in areas of the apical membrane consistent with their presence in stellate cells of the *Drosophila* Malpighian tubule (O'Donnell *et al.*, 1998). Both fluid secretion and the calcium-mediated increase in anion conductance are inhibited by Cl^- channel blockers such as niflumic acid. Vibrating probe analysis has revealed a small number of current density hot spots, coincident with stellate cells, and currents associated with the stellate cells are reduced by low-chloride saline or chloride channel blockers. Targeted expression of an aequorin transgene has been used to demonstrate that the neurohormone leucokinin elicits a rapid increase in intracellular calcium levels in stellate cells before physiological effects are evident. In sum, these data show that leucokinins act on *Drosophila* tubule stellate cells to elevate intracellular calcium and thereby increase transcellular chloride conductance through channels.

For tubules of both *Drosophila* and *Aedes* secreting at the basal rate there is agreement that transepithelial Cl^- flux is mediated by stellate cell Cl^- channels (O'Donnell *et al.*, 1998; O'Connor and Beyenbach, 2001). In *Aedes* tubules, however, a paracellular pathway has been proposed in response to stimulation with leucokinin. The transepithelial resistance of *Aedes* tubules falls 4.3-fold in response to leucokinin, whereas the input resistance of the principal cells falls only 1.7-fold (Masia *et al.*, 2000). Coupled with the finding that leucokinin reduces transepithelial resistance in specific regions of tubules apparently lacking stellate cells (Yu and Beyenbach, 2004), this suggests that leucokinin acts on a site separate from both cell types, i.e. the paracellular pathway. In support of this finding, the large transepithelial Cl^- diffusion potentials induced by leucokinin are similar for lumen-to-bath and bath-to-lumen transepithelial gradients.

The latter finding is most easily explained on the basis of a single diffusion barrier such as a septate junction.

3.3 AQUAPORINS AND OSMOTIC PERMEABILITY OF MALPIGHIAN TUBULES

The osmotic permeability of the *Rhodnius* Malpighian tubule wall has been measured by simultaneous perfusion of internal and external surfaces. The measured permeability of the upper tubule, 4.3×10^{-3} cm s^{-1} $osmol^{-1}$, can be increased 35%, to 5.8×10^{-3} cm s^{-1} osmole, when corrections are made for unstirred layer effects (O'Donnell *et al.*, 1982). The high osmotic permeability of the tubule wall is one factor which indicates that the route for transepithelial water movement during fluid secretion is transcellular. A transcellular route, rather than a paracellular one, is also more likely given the very small frontal area of the intercellular clefts. The paracellular clefts between *Rhodnius* tubule cells are about 17 nm in width and occupy about 0.039% of the frontal area of the tubules. If all the fluid moved through the clefts (120 nl min^{-1}) it would pass through at 660 μm $s^{-1}$and the fluid would take only 22 ms, therefore, to traverse the 15 μm length of the cleft. Small ions and solutes such as urea take 2–3 times as long to diffuse 15 μm, and larger solutes such as sucrose would take longer still. One would therefore expect a strong correlation between the rate of fluid transport and passive transepithelial movements of extracellular markers. In particular, entrainment of solutes into a rapidly flowing fluid will accelerate transepithelial solute movement. However, neither sucrose nor inulin crosses the wall of the tubule any faster during rapid fluid secretion (Maddrell, 1981). This absence of significant solute/solvent coupling in the clefts, together with the small number and area of the clefts, militates against significant paracellular water flow during secretion by the tubules.

Measurements by Whittembury *et al.* (1986) raised the possibility of some paracellular fluid flow, based on entrainment of some solutes (urea, mannitol, sucrose) but not others (inulin, raffinose, dextran) in fluid flow. Interpretation of these findings is complicated by the finding that changes in intraluminal hydrostatic pressure can greatly alter the rates of permeation of molecules such as sucrose (O'Donnell *et al.*, 1983).

The most convincing argument against paracellular flow and in support of transcellular flow is the discovery of aquaporins (water channels) in Malpighian tubules and the finding that tubule osmotic permeability is reduced by mercurial agents which block aquaporins (Hernandez *et al.*, 2001; Echevarria *et al.*, 2001). Expression of the mRNA for the *Rhodnius* aquaporin is increased after a blood meal and in tubules treated with 5-hydroxytryptamine or cAMP (Martini *et al.*, 2004). In the tubules of the cricket *Acheta domesticus*, aquaporin expression appears to be hormonally regulated in both apical and basolateral membranes (Spring *et al.*, 2007). It thus appears that transporters used for the excretion of excess water and excess ions can be regulated in insect Malpighian tubules. Aquaporins are also expressed in both the principal cells and stellate cells of the *Drosophila* Malpighian tubule (Kaufmann *et al.*, 2005). One of the aquaporins, *Drosophila* integral protein (DrIP) is similar to human AQP4, a channel that exhibits the highest rates of water transport, and it is localized to the stellate cells. DrIP contains several residues consistent with its actions as a water-specific channel.

4 Secretion and reabsorption by downstream segments of the Malpighian tubule and the hindgut

A common feature of the insect excretory system is the modification of the fluid secreted by Malpighian tubules in a downstream segment of the tubules themselves or in the hindgut. Although reabsorption of useful ions and water is accomplished by the hindgut in most species, in some cases a proximal (lower) segment of the Malpighian tubule is involved.

4.1 REABSORPTION BY THE MALPIGHIAN TUBULES

For the blood feeder *R. prolixus*, the high surface area offered by the lower Malpighian tubule relative to the volume of fluid in its lumen permits rapid reabsorption of KCl during the post-prandial diuresis. In *Drosophila*, reabsorption of both K^+ and water in the lower Malpighian tubule probably serves a preparatory role, prior to additional reabsorption in the hindgut. The fluid secreted by the upper Malpighian tubule of *Rhodnius* (approximately 100 mM NaCl, 80 mM KCl) is modified by reabsorption of almost all the KCl but not water as the secreted fluid passes through the 30% of the lower tubule's length closest to the hindgut. Osmotic permeability of the lower tubule is much lower in this region relative to the upper tubule or the upper region of the lower tubule (O'Donnell *et al.*, 1982). Water is thus not reabsorbed as a consequence of KCl reabsorption, so what passes into the hindgut is both NaCl-rich and hypoosmotic. *Rhodnius* feeds on blood which is hypoosmotic to its own hemolymph, so this process of isosmotic secretion by the upper tubule followed by KCl reabsorption by the lower tubule preserves osmotic and ionic homeostasis.

The transporters implicated in reabsorption of K^+ and Cl^- by the lower Malpighian tubule differ fundamentally from those which secrete these ions across the upper Malpighian tubule (Fig. 8). The current model proposes that K^+ is moved thermodynamically uphill from lumen to cell by an ATP-dependent pump, resembling the H^+/K^+-ATPase of the gastric mucosa, and that K^+ moves passively from cell to bathing saline (hemolymph) through an electrodiffusive pathway, i.e. K^+ channels. K^+ reabsorption is unaffected by amiloride (which blocks a putative K^+/H^+

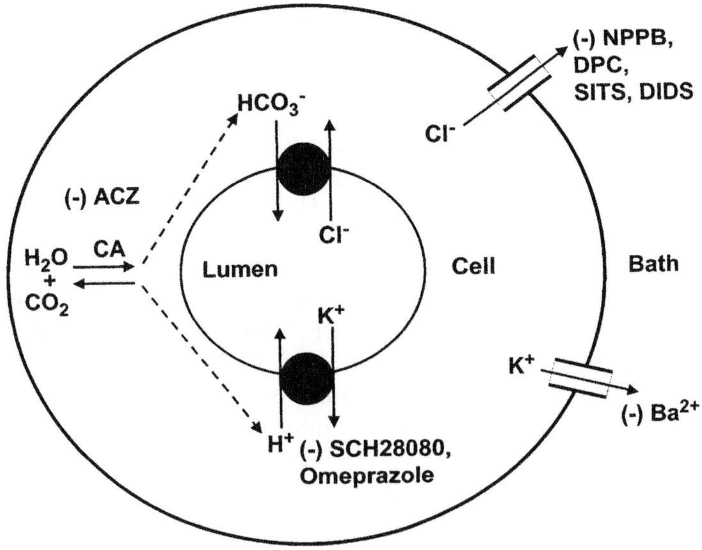

FIG. 8 Schematic diagram summarizing the working hypothesis for the mechanisms of KCl reabsorption by the lower Malpighian tubule of *Rhodnius prolixus*. On the apical membrane, an H^+/K^+-ATPase inhibitable by drugs such as omeprazole and SCH 28080 exchanges lumenal K^+ for cellular H^+ (Haley and O'Donnell, 1997). Lumenal acidification is prevented by a compensatory flux of base through a putative SITS-insensitive Cl^-/HCO_3^- exchanger. Cl^- exits the cell through channels which can be blocked by diphenylamine-2-carboxylate (DPC) and 5-nitro-2-(3-phenylpropylamino) benzoic acid (NPPB), 4-acetamido-4′-isothiocyanatostilbene-2,2′- disulfonic acid (SITS) and diisothiocyanatostilbene-2,2′-disulfonic acid (DIDS). K^+ exits the cell through Ba^{2+}-sensitive channels (Haley and O'Donnell, 1997). Acetazolamide (ACZ) inhibits carbonic anhydrase (CA), thereby inhibiting KCl reabsorption by preventing formation of HCO_3^- and H^+ for the apical exchangers. Redrawn from Haley *et al.* (1997, Fig. 5). Reprinted with permission of Elsevier Science.

exchanger in the upper tubule) and by the V-type-ATPase inhibitor bafilomycin A_1, applied to the basolateral or apical services of the lower tubule (Haley and O'Donnell, 1997). By contrast, K^+ reabsorption is inhibited by drugs such as omeprazole, SCH 28080, SKF 96067 and SKF 96356. These drugs are either activated by low pH or are weak bases which tend to accumulate in acidic compartments. The pH of fluid in the lower tubule is 8.48, approximately 1.2 units more alkaline than the fluid secreted by the upper tubule. The concentration of omeprazole and the other drugs which inhibit lower tubule KCl reabsorption are higher, therefore, than those required to inhibit gastric H^+ secretion in vertebrates, but comparable to the levels which inhibit H^+/K^+-ATPases in other tissues (Haley and O'Donnell, 1997). K^+ reabsorption by the lower tubule is also blocked by the K^+ channel blocker Ba^{2+}. Changes in basolateral membrane potential in response to changes in bathing saline K^+ concentration and blockade of these changes by Ba^{2+} also indicate the presence of basolateral K^+ channels.

The current model for the mechanism of chloride reabsorption by the lower tubule proposes that chloride moves from lumen to cell through a stilbene-insensitive Cl^-/HCO_3^- exchanger and then exits the cell through basolateral chloride channels (Fig. 8; Haley *et al.*, 1997). Changes in basolateral membrane potential and transepithelial potential in response to bathing saline Cl^- concentration indicate the presence of Cl^- channels in the basolateral membrane. The blockade of Cl^--dependent changes in potential and KCl reabsorption by Cl^- channel blockers but not by compounds that inhibit K^+:Cl^- and Na^+:K^+:$2Cl^-$ cotransporters are also indicative of a role for chloride channels. Inhibition of KCl reabsorption by carbonic anhydrase inhibitors such as acetazolamide suggests that cellular CO_2 is a source of both H^+ and HCO_3^- for exchange of K^+ and Cl^-, respectively. Some H^+ and HCO_3^- may combine in the tubule lumen to form CO_2, which may diffuse back across the tubule ball into the bathing saline. The pH of the secreted fluid will thus be set by the rates of both H^+ and HCO_3^- influx and CO_2 loss.

The lower tubule of *D. melanogaster* reabsorbs fluid and transports ions into or out of the lumen (O'Donnell and Maddrell, 1995). Secretion of Ca^{2+} by the lower tubule and acidification of the lumen are discussed in Section 6.1. The K^+-rich fluid secreted by the upper tubule is modified by reabsorption of K^+ and water. Modification of the urine by reabsorption is also accomplished by the lower tubule or ampulla or both in the house cricket *Ac. domesticus* (Spring and Hazelton, 1987), and possibly by reabsorptive cells scattered throughout the epithelia in tubules of the cricket *Te. oceanicus* (Marshall *et al.*, 1993). Reabsorption by specific cells or regions of the Malpighian tubule thus reduces the amount water and ions that must be reabsorbed by the hindgut. A similar role is performed by the ileum of the locust, *S. gregaria* (Phillips *et al.*, 1986). Reabsorption of water and ions in the ileum allows the rectum, downstream of the ileum, to process more thoroughly a reduced volume of fluid delivered to it.

4.2 THE INSECT HINDGUT

The hindgut, and the rectum, in particular, is a major site of water conservation in insects. The hindgut modifies the fluid secreted by the Malpighian tubules by reabsorbing useful solutes and water and eliminating those molecules which are present in excess or are toxic. Up to 80–95% of the ions and water secreted by the Malpighian tubules is reabsorbed in the hindgut of stick insects (Ramsay, 1955a,b), water-stressed locusts and cockroaches (Phillips 1964a,b; Wall, 1971), and about 65% in blowflies. In insects such as the mealworm larvae of

Tenebrio molitor water reabsorption in the rectum is so efficient that the fecal pellets which are expelled are almost completely dry. Water reabsorption in the insect hindgut is tightly regulated. For example, less than 10% of water is reabsorbed from the hindgut when locusts are fed on a succulent diet (Loveridge, 1975), and blood-feeders such as *Rhodnius* recover virtually no water during diuresis.

Four distinct physiological mechanisms of water conservation are seen in the hindgut of different species (Bradley, 1985; Phillips *et al.*, 1986):

(1) *absorption of hypoosmotic fluid by rectal pads or papillae in cockroaches, locusts and adult blowflies*. The classic paper by Wall (1971) demonstrated the importance of local osmosis and solute recycling during fluid absorption. Active transport of Na^+, K^+ and Cl^- creates the osmotic gradients for absorption of water from the rectal lumen. Fluid and ions move from the lumen into the rectal epithelial cells and then into the intercellular spaces between the epithelial cells. As the solutes and water move along the intercellular channels, the ions are reabsorbed into the epithelial cells for subsequent return to the intercellular spaces so as to permit further osmotic reabsorption from the lumen. Microsampling of fluid from the rectal tissues provides direct support for this scheme (Wall, 1971). Fluid in the lateral intercellular spaces near the lumen is hyperosmotic, and the total ion concentration in the intercellular channel declines toward the hemocoel.

The elegant work of Phillips and co-workers suggests that in the locust rectum a primary mechanism of Cl^- transport is responsible for absorption of ions and water (Phillips *et al.*, 1986, 1996). This Cl^- transport is not coupled to or driven secondarily by movements of Na^+, K^+, HCO_3^-, Ca^{2+} or Mg^{2+}. Although an apical V-type H^+-ATPase acidifies the hindgut lumen, it does so at a rate that is only 10–15% of Cl-dependent short-circuit current across the rectum. Chloride transport is stimulated by chloride transport stimulating hormone (CTSH) and ion transport peptide (ITP), both of which act through intracellular cAMP.

S. gregaria ITP (SchgrITP) is the first reported insect member of a large family of crustacean neuropeptide hormones that control blood sugar levels, molting and reproduction. A partial amino acid sequence of ITP purified from locust nervous corpus cardiacum was used to derive degenerate primers that were used to amplify a cDNA from brain RNA using reverse transcription and the polymerase chain reaction (RT-PCR). The cDNA sequence was extended using anchored PCR to yield a partial, 517bp cDNA clone which encodes a putative ITP prohormone (Meredith *et al.*, 1996). The prohormone can be cleaved at two dibasic amino acid sites to yield a 72 residue active amidated peptide. The deduced amino acid sequence from the cDNA agrees completely with the amino acid sequence and molecular mass (8564 Da) derived from analysis of purified ITP. Native ITP has the same actions as crude CC extracts in stimulating Cl^-, Na^+, K^+ and fluid absorption. *Drosophila* Kc1 cells transfected with preproITP secrete a peptide (KcITP) that is very active in the locust ileal bioassay (Wang *et al.*, 2000; Zhao *et al.*, 2005). Site-directed mutagenesis and expression of KcITP mutants has shown that that amidated leucine (residue 72) at the C-terminus is essential for receptor stimulation.

(2) *secretion of hyperosmotic salt solutions by salt glands.* Salt water mosquito larvae which are capable of hypoosmotic regulation produce a concentrated urine by the secretion of a hyperosmotic fluid across the posterior rectum. Reabsorption of ions and useful metabolites may occur across the anterior rectum (Strange and Phillips, 1985; Bradley, 1987). Na^+, K^+, Mg^{2+} and Cl^- are all actively transported by the posterior segment to the lumen side (Bradley and Phillips, 1977).

(3) *secretion of hyperosmotic salt solutions within the cryptonephridial complexes of beetle larvae.* The distal portions of the Malpighian tubules of *Te. molitor* are held closely to the surface of the rectum by an enveloping perinephric membrane, so that the rectum and the tubules act in concert. The tubules secrete K^+, Cl^- and to a lesser extent Na^+ into the lumen. The concentration of KCl in the tubule lumen in some tenebrionid larvae approaches and may even transiently exceed saturation. The resultant 'osmotic sink' (Machin, 1983) is used for fecal dehydration or atmospheric water vapour absorption. The ultimate source of ions for transport into the tubule lumen is the hemolymph. The perinephric space is a functional compartment beneath the perinephric membrane and surrounding the tubules. Potassium activities in the perinephric space are close to electrochemical equilibrium with the hemolymph, whereas the activities of Na^+ and H^+ are reduced fivefold and threefold, respectively, below the corresponding Nernst equilibrium values. It appears that cations move from hemolymph across the perinephric membrane into the perinephric space, which is the immediate source for cations transported into the tubule cells and then the lumen (O'Donnell and Machin, 1991). The main function of the perinephric membrane is to provide an osmotic barrier so water does not move from the hemolymph into the complex in response to the osmotic sink within the Malpighian tubule lumen; instead, water is extracted from the rectal lumen.

(4) the use of anal sacs in thysanurans (silverfish and firebrats) and flea larvae. In this case, there is no evidence for high concentrations of

inorganic salts (O'Donnell and Machin, 1988).The anal sac is formed from a highly folded single-layered epithelium and the subcuticular space between the rectal cuticle and the epithelium is filled with a material that resembles glycerol and also contains mucopolysaccharides. Dehydration of this material by transport activity of the epithelial cells creates the necessary osmotic gradient to drive water reabsorption from the lumen. One model of water transport proposes that water movment is produced by electro-osmosis (Kuppers *et al.*, 1986). Active transport of K^+ across the apical membrane of the rectal epithelial cells generates a lumen-positive potential of 200 mV. Cations then return to the cells through cation-selective channels, and water is couple electro-osmotically to the inward current. Water thus moves from the rectal lumen onto the hygroscopic material of the subcuticular space and is then removed to the cytoplasm of the epithelial cells. This proposal has been criticised on several technical grounds, foremost of which is that the proposed coupling of ions to water occurs in the subcuticular space at water activities which are well below the saturation point of KCl (O'Donnell and Machin, 1988).

5 Nitrogenous waste excretion

Ammonia resulting from catabolism of nitrogen-rich compounds such as amino acids is toxic and must be either eliminated from the body or converted into less toxic forms. Whereas aquatic species typically excrete ammonia as the predominant nitrogenous waste, many terrestrial insects expend metabolic energy to synthesize uric acid. Some larger species, or those with access to water, excrete ammonia, either alone or in combination with uric acid.

5.1 URIC ACID

5.1.1 *Synthesis and storage of uric acid*

Two properties suit uric acid for the purpose of nitrogen excretion: it contains 33% nitrogen, and its solubility is particularly low (1.1 mM at pH 5.5) under the acid conditions found in the terminal section of the excretory system (Phillips, 1977). Elimination of uric acid or urate salts thus leads to the excretion of large amounts of nitrogen without incurring much loss of water. Uric acid is synthesized primarily in the fat body. During larval growth, uric acid is released into the hemolymph shortly after synthesis and is subsequently excreted by the Malpighian tubules.

Xanthine oxidase, which catalyses uric acid formation from its immediate precursor xanthine, is present in fat body, Malpighian tubules and intestine in the silkworm *Bombyx mori* (Hayashi, 1960; Komoto *et al.*, 1999). In *Drosophila*, a search for xanthine oxidase in FlyAtlas: the *Drosophila* adult gene expression atlas (Chintapalli *et al.*, 2007), shows that the genes are much more enriched in the tubules than in the gut or fat body.

The excretory system is non-functional during metamorphosis of holometabolous insects. In lepidopterans, therefore, uric acid is sequestered in the fat body during larval-pupal metamorphosis (Buckner, 1982). In fly pupae, uric acid is accumulated not only in the fat body, but also in the intestine and the Malpighian tubules in response to the shut down of the excretory organs during this phase of the life cycle (Schwantes, 1990). The switch from excretion to storage is under hormonal control in larval *Ma. sexta* and it occurs during the transition from the feeding to the wandering stage (Buckner, 1982). There is an increase in fat body urate levels and a corresponding decline in hemolymph urate concentration at this time. An increase in the titre of 20-hydroxyecdysone and a decrease in the titre of juvenile hormone I control the switch from uric acid excretion to storage in the fat body. Fat body cells respond to 20-hydroxyecdysone and switch from secretion to storage of urate only in the absence of juvenile hormone. When exogenous juvenile hormone I or the analog methoprene are injected, storage is prevented or reversed and there is no decline in hemolymph uric acid levels.

During the subsequent transformation from pupa to adult, the fat body releases stored urate, and the timing of release corresponds to declining levels of 20-hydroxyecdysone in the hemolymph (Ehresmann *et al.*, 1990). Release of uric acid can be delayed by exogenous 20-hydroxyecdysone, but not during the final stages of transformation. A second ancillary hormone may be involved, therefore.

Urate is stored in the fat body of *Ma. sexta* pupae as discrete membrane bound vacuoles, termed uric acid storage vacuoles, which can be stained with reduced silver to reveal stored uric acid. Electron microscopy shows the vacuoles to be spherical or biconcave, and 0.5–2 μm in diameter (Buckner *et al.*, 1985). Biochemical and elemental analyses indicate that the storage vacuoles contain 75% uric acid, 5% protein, 0.5% carbohydrate, 6–7% water, 6.7% potassium and 0.5% sodium. At high magnification, the granules appear in micrographs as concentric whorls of fibrous or filamentous material. Granules appear to be constructed from tightly coiled fibres of crystalline uric acid or urate salts, and each fibre is enveloped with protein (Buckner *et al.*, 1990).

Uric acid is also stored in wings or larval cuticle in Lepidoptera, presumably to provide pigmentation. Levels of uric acid in the hemolymph increase during development of *Ma. sexta* pupae, possibly as a consequence of the release of uric acid from epidermal stores

(Buckner and Newman, 1990). The uric acid is subsequently either excreted or stored in the fat body. Uric acid is also stored in the eyes in *Drosophila*; this storage excretion allows the crystals to act as reflectors to help light penetrate more deeply into the eye. In flies with the mutation *rosy* the eyes are dull red, because they lack uric acid crystals around their ommatidia (Reaume *et al.*, 1991).

5.1.2 *Active transport of urate by Malpighian tubules*

Although insects fed a nitrogen-poor diet may retain uric acid internally, those with a protein-rich diet require mechanisms for elimination of urates or uric acid. In *Rhodnius* urate is excreted at a rate of up to 1.2 mg day^{-1}, more than 15% of the basic body weight per day (O'Donnell *et al.*, 1983). The tsetse fly, *Glossina morsitans* Westw., must excrete nearly 50% of the dry weight of the ingested blood in the form of nitrogen-containing compounds (mainly uric acid, but also arginine and histidine) merely to dispose of surplus nitrogen (Bursell, 1965). It is important to point out that the conclusion of active transport is based on measured concentrations of urate and in hemolymph and tubule lumen and on the transepithelial potential. However, if urate is immediately converted to uric acid and precipitated, a gradient favouring passive movement of urate into the lumen may be created.

Although uric acid is now known to be transported actively by the lower Malpighian tubule of *Rhodnius* (O'Donnell *et al.*, 1983), a 'secretion-reabsorption' model for passive uric acid transport in the upper Malpighian tubules of *Rhodnius* was first proposed by Wigglesworth (1931), and is reproduced in many current textbooks of entomology or comparative physiology. Wigglesworth proposed that urate moved passively across the cells and into the lumen of the upper secretory portion of the tubules, and was then carried by entrainment in the flow of fluid into the lower tubule, where precipitation of uric acid occurred as a result of acidification and water reabsorption. The model proposed that acidification resulted from reabsorption of bicarbonate. Wigglesworth's model was subsequently modified by Miles (1966), who suggested that acidification may be a direct result from anisotropy of oxidation-reduction during electron transport in the cells of the excretory system.

A requirement of the models of both Wigglesworth and Miles is a significant flux of urate across the upper tubule. However, measurements of the rate of passive movement of urate across the *upper* fluid secreting segment of the tubule indicate that this segment of the tubule accounts for less than 2.3% of the uric acid eliminated by the insect. Measurements with isolated tubules demonstrate that urate is actively transported against an electrochemical gradient into the lumen of the *lower* Malpighian tubule (O'Donnell *et al.*, 1983). The lower 1/3 of the lower tubule is also responsible for reabsorption of K^+ and Cl^- but not water. Subsequent histochemical studies by Wigglesworth confirmed the exclusive role of the lower tubule in urate transport (Wigglesworth, 1987a,b). Wigglesworth notes that he had favoured the 'secretion/reabsorption' hypothesis in order to accommodate the low solubility of uric acid combined with the massive excretion of it. However, his histochemical studies some 46 years after his initial study revealed that urates or uric acid pass through the basal lamina and the cytoplasm of the cells of the lower segment in high concentration (presumably in association with protein).

Urate transport has been identified in the isolated Malpighian tubules of locusts, butterfly larvae and mantids, as well as *Rhodnius*, suggesting that the mechanism is widespread (O'Donnell *et al.*, 1983). Although most studies have dealt with nymphs or adult insects, uric acid deposition in the tubule lumen has been demonstrated in the embryo in *Drosophila*, prior to emergence of the first instar larva (Ainsworth *et al.*, 2000). In *Calliphora*, urate oxidase activity in the Malpighian tubules converts uric acid to a related nitrogenous compound, allantoin, which is then excreted. Urate oxidase activity is not detectable during embryonic development of *Drosophila*, in first and second instar larvae, or in pupae, but is present in third instar larvae and adults.Through control by a developmental clock within the Malpighian tubules, urate oxidase activity in the tubules reappears at the time of emergence of the imago from the puparium (Friedman and Johnson, 1977). Dow and Davies (2003) have included a section on purine metabolism in *Drosophila* in their comprehensive review. They note that the presence of allantoinase in third instar larvae and adults implies that urea could be produced. However, the flies lack a functional urease, so production of ammonia from any urea via this route seems unlikely.

In contrast to the mechanism of urate transport in vertebrates, transport by isolated tubules is ouabain-insensitive and results in precipitation of free uric acid rather than urate salts (O'Donnell *et al.*, 1983). *In vivo*, higher urate transport rates are induced in *Rhodnius* by increases in hemolymph urate concentration as the blood meal is digested in the days following feeding (Fig. 9). A similar induction of transport of organic anions such as para-aminohippurate after the blood meal has been reported by Maddrell and Gardiner (1975). *In vitro* experiments with isolated tubules of the tsetse fly suggest that induction of urate transport may be an autonomous response of the tubule cells to increases in bathing fluid urate concentration, and neurohormonal mediation of the response may not be not required (O'Donnell *et al.*, 1983).

In addition to entrainment in fluid movement down the lower Malpighian tubule of *Rhodnius*, uric acid crystals are also moved as a consequence of the beating of axopods (Bradley and Satir, 1979), structures more commonly found in protozoans. The axopods arise from

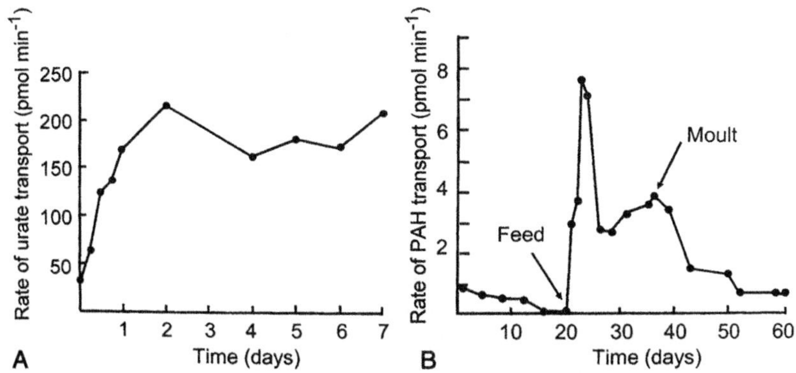

FIG. 9 Induction of (A) urate secretion and (B) para-aminohippuric acid (PAH) in *Rhodnius* Malpighian tubules in response to feeding. (A) Urate secretion rates (mean ± S.E.M) for the indicated number of 5th instar tubules were measured *in vitro* in saline containing 0.5 mM urate. Animals were fed a blood meal on day 0. Redrawn from O'Donnell *et al.* (1983, Fig. 6). (B) PAH transport in Malpighian tubules isolated from 4th stage larvae of *Rhodnius* starved until day 10 then fed and allowed to moult to the 5th instar on day 21. Tubules were bathed in saline containing 0.05 mM tritiated PAH. Redrawn from Maddrell and Gardiner (1975, Fig. 1). Reprinted with permission from the Company of Biologists.

the luminal cell membrane and are 0.2–0.8 μm in diameter and 10 μm or more in length. Each contains up to 46 microtubules. The axopods of the *Rhodnius* lower Malpighian tubule differ from those in protozoa in that they branch as they extend outward from the luminal cell surface. The beating of the axopods moves urate crystals, as observed in short lengths of lower tubules, isolated *in vitro* and open at both ends. Tubules isolated in this way are unaffected by fluid movements produced by peristalsis in the rectum or fluid secretion by the upper tubule. The lower tubule itself is reabsorptive rather than secretory, and time lapse photography studies by Bradley and Satir (1979) show an absence of peristalsis during movement of urate crystals in the lumen. Crystal movement has been shown to be restricted to the open lumen, near the tips of the axopods, and does not occur at the periphery of the lumen, near microvilli.

The need for movement of urate crystals by axopods is apparent when one considers the other functions of the lower tubule of *Rhodnius*. This segment reabsorbs K^+ and Cl^-, but not water. It has thus taken over the function of the hindgut in many other insects. However, the same characteristics which suit the lower tubule for KCl reabsorption, such as a narrow lumen, extensive microvilli and a high-surface area to volume ratio, create a potential for blockage by urate crystals, particularly given the low flow rates of water when the blood meal is digested 3–4 days after diuresis and the high rates of urate transport at this time. Movement of uric acid crystals down the lower tubule will thus tend to minimize the tendency for the tubule to become blocked. Such translocation is normally accomplished by ciliated epithelia in other animal groups. Cilia are not found in the somatic cells of arthropods, so it has been suggested that the axopods of the *Rhodnius* lower tubule may represent an example of convergent evolution (Bradley and Satir, 1979).

Although the Malpighian tubules provide a potent means for elimination of excess urate, some species retain high levels of this molecule because it may serve other purposes in the insect's biology, in addition to its role as a nitrogenous waste. In *Rhodnius*, high levels of urate in the hemolymph acts as an antioxidant, as described below in Section 7.3. Cockroaches utilize stored urate as part of an osmoregulatory/ionoregulatory strategy. Urate levels in the fat body increase when cockroaches are dehydrated or starved. Urate appears to be stored primarily as salts of Na^+, but also K^+ and possibly NH_4^+ when concentrations of these ions increase, during dehydration. These ions can then be released during rehydration. Urate appears to be present in fat body 'spherules', formed as concentric laminations and typically about 9–11 μm in diameter (Tucker, 1977a). Urate and K^+ are mobilized from the spherules during dehydration, whereas sodium urate deposits increase at the same time (Tucker 1977a,b). Sodium urate may be present as amorphous granules as well as spherules. It is suggested that uptake of Na^+ by the fat body enables the animal to maintain a fairly constant level of Na^+ in the hemolymph during dehydration, without having to excrete large amounts of Na^+.

Cochran (1985) has championed the idea that uric acid plays a central role in cockroach physiology. In addition to the ionoregulatory role for urates in maintenance of hemolymph Na^+ levels, urate excretion plays a regulatory role in the nitrogen economy of cockroaches, with excretion in *Parcoblatta fulvescens* occurring only when excess nitrogen is available in the diet. In most cockroaches, urates are stored in the fat body when the animals are fed nitrogen-rich diets, and are mobilized when the insects are fed carbohydrate-rich diets (Mullins and Cochran, 1975). Animals fed ^{14}C-labelled hypoxanthine incorporate the label into uric acid, and animals on low nitrogen diets release $^{14}CO_2$ to the atmosphere, consistent with degradation of uric acid. Degradation appears to involve microbial symbionts in the fat body cells. Although the majority of cockroaches excrete ammonia rather than uric acid, as described below, wood cockroaches and a few other species in sub-families Blatellinae and Plecopterinae void pellets which are high in urates, and urate spherules are present in the lumen of the Malpighian tubules of these species (Lembke and Cochran, 1988). The spherules form an opaque white slurry in the tubule lumen. It is suggested that tubules of all cockroaches can transport urate, but that it is bound to hemolymph protein (Hopkins and Lofgren, 1968) in those species which do not void it. Birefringent vacuoles within the tubule cells may indicate vesicular transport of urate, although other materials could be the source of the birefringence (Lembke and Cochran, 1988).

5.2 AMMONIA TRANSPORT IN INSECTS

5.2.1 *Aquatic species*

Although most terrestrial insects are uricotelic, classic work by Staddon (1955, 1959) showed that aquatic species have reverted to ammoniotelism. This finding is in keeping with the original proposal of Delaunay (1931) that synthesis of more complex molecules from ammonia by terrestrial invertebrates is essentially a detoxification mechanism necessitated by a restricted water supply. Aquatic larval stages of insects such as lacewings (Neuroptera; Staddon, 1955) and dragonflies (Staddon, 1959) excrete ammonia as the primary nitrogenous waste. Lacewing larvae excrete 90% of their total nitrogenous waste as ammonia, and secretion of ammonium by the Malpighian tubules is suggested (Staddon, 1955). The latter proposal is in contrast with measurements in locusts (see below), although it is well known that Malpighian tubules of some terrestrial insects can secrete NH_4^+ in place of K^+ (Maddrell, 1969). In dragonfly larvae there is a transient increase in ammonia excretion following a protein-rich meal, whereas is there is no change in the output of uric acid (Staddon, 1959). At least 60% of the nitrogen absorbed by the larvae within 24–48 h of feeding is excreted as ammonia during the same period.

The aquatic larvae of the mosquito *Ae. aegypti* also excrete NH_4^+, and in this species an important site of excretion is the anal papilla (Donini and O'Donnell, 2005; Donini *et al.*, 2006). The four papillae arise from an extension of the terminal segment and project into the external medium. Studies using the scanning ion-selective electrode technique have shown that the papillae take up Na^+, K^+ and Cl^- from the water and that they also excrete H^+ and NH_4^+. Although the mechanism of NH_4^+ transport is unknown, the data are consistent with a role for diffusion trapping by protons. Earlier studies of whole larvae using radioisotopes did not show an effect of NH_4^+ on Na^+ influx, suggesting that Na^+/NH_4^+ exchange is not involved (Stobbart, 1967). NH_4^+ flux across the papilla in fresh water at pH 7.4 is 360 nmol cm^{-2} h^{-1}, similar in magnitude to fluxes across the isolated locust hindgut (580 nmol cm^{-2} h^{-1}; Thomson *et al.*, 1988) and fluxes of 220 nmol cm^{-2} h^{-1} calculated by the latter authors using whole animal data of Mullins (1974) for *Periplaneta americana*. Possible involvement of a two ATPases in driving ion transport by the papillae is suggested by the work of Patrick *et al.* (2006). Immunohistochemical labelling shows that P-type Na^+/K^+-ATPase and V-type H^+-ATPase are localized to the basal and apical surfaces, respectively, of the anal papillae.

5.2.2 *Terrestrial species*

There also appear to be numerous exceptions to the generalization that terrestrial insects are primarily uricotelic. Studies of cockroaches (Mullins and Cochran, 1972, 1973a) and locusts (Thomson *et al.*, 1988) indicate an important role for ammonia excretion. Up to 90% of the nitrogenous material in the excreta of cockroaches such as *Periplaneta* is in the form of ammonia and little or no uric acid is detectable. Amino acids are also found in the excreta, possibly as a result of microbial activities in the gut or incomplete absorption from food. Ammonia is also a major excretory product in some lepidopteran larvae as well (Kuzhivelil and Mohamed, 1987), although the possible contribution of gut microbial activity was not ruled out in the latter study. Urea can also be a major excretory product in some lepidopteran larvae (Lazar and Mohamed, 1979). Detailed studies of desert locusts (*S. gregaria*) have revealed excretion of ammonium at three times the rate of total urate and ammonium concentrations in the excreta as high as 400 mM (Harrison and Phillips, 1992). Ammonium excretion is compatible with water conservation in this species because ammonium is present in precipitated form as ammonium urate. Harrison and Phillips (1992) point out that many early studies of nitrogenous excretion in insects may have underestimated the importance of ammonium excretion because pellets were either oven-dried, resulting in volatilization and loss of NH_3, and/or dilution levels during preparation of pellets for analysis were insufficient to release ammonium from precipitated form.

The majority (> 50%) of NH_4^+ is excreted across the ileum in locusts; less than 10% of NH_4^+ excretion is accounted for by the Malpighian tubules (Stagg *et al.*, 1991). The ileum performs several important and interlinked homeostatic functions in locusts (Peach and Phillips, 1991). Amino acids such as proline are reabsorbed from the gut lumen and metabolized to produce ammonia. Cellular NH_4^+ is then exchanged for Na^+ in the lumen, thus driving reabsorption of Na^+, which is in short supply in the herbivorous diet. This exchange is suggested by the finding that the rate of transport of ammonia, but not H^+, is directly affected by Na^+ substitution or amiloride. The ileum is also a major site of acid excretion, in the form of H^+ trapped by NH_3 as NH_4^+. However, ammonia transport is independent of transmembrane pH gradients and luminal proton secretion. These findings thus rule out the 'diffusion trapping' mechanism of ammonia excretion typical of mammalian transporting epithelia (e.g. Knepper *et al.*, 1984). Instead, it appears that ammonia crosses the apical membrane of the locust ileum via an amiloride-sensitive Na^+/NH_4^+ exchange mechanism (Thomson *et al.*, 1988). Whereas proline is absorbed from either the lumen or the hemolymph in the ileum, it must be supplied to the luminal side in the rectum. In the latter tissue, proline transport exceeds the metabolic requirements of chloride transport more than 10-fold, and appears to function not just as an energy source but also as an important osmolyte in fluid reabsorption (Chamberlin and Phillips, 1982), as described in Section 5.3. Proline also functions as an important osmolyte in mosquito larvae (Section 5.3).

Larvae of blowflies and flesh flies are commonly faced with high concentrations of ammonia as they burrow into and feed upon rotting flesh or living tissue. Larvae of the flesh fly *Sarcophaga bullata* ingest ammonia, reabsorb it from the midgut, and then excrete it across the hindgut epithelium. In addition, they are exposed to excess ammonia resulting from protein metabolism. Deamination of amino acids and active secretion of NH_4^+ have been studied in isolated hindgut preparations of blowflies by Prusch (1972, 1976). Ammonia can be secreted against a pH gradient (lumen alkaline) which would not be favourable for NH_3 diffusion. Both K^+ and Cl^- are actively secreted into the lumen, and the mechanism of NH_4^+ secretion is independent of that for K^+. The mechanism of NH_4^+ secretion differs from the process of Na^+/NH_4^+ exchange in the locust ileum. Na^+ is passively distributed between the hindgut lumen and hemolymph, accumulating in the lumen in response to the lumen negative transepithelial potential of about -15 mV. NH_4^+, K^+ and Na^+ can be concentrated in the lumen simultaneously, so exchange of NH_4^+ with K^+ or Na^+ seems unlikely. However, ammonium secretion is reduced by about 60% when either K^+ or Na^+ is removed, suggesting that both alkali cations are implicated in some way in NH_4^+ secretion.

The larvae of sheep blowflies are also exposed to high concentrations of ammonia. The larvae live in urine-soaked wool, proteinaceous sera and blood from the lesions produced by the larvae themselves. The larvae are exposed to excess nitrogen resulting not only from metabolism of proteins, but from the urine soaking the wool around the lesions. The larval environment fluctuates widely in ionic and osmotic composition, since sheep urine varies from 800 to 2500 mOsm kg^{-1} and osmotic pressures of fluids in the lesions will rise above the plasma value of 290 mOsm kg^{-1} as wounds dry out. Studies of ionic and osmotic regulation by larvae of the sheep blowfly *Lucilia cuprina* show that NH_4^+ concentrations in the hemolymph are not altered significantly from the very high control value (16 mM) when the concentration of NH_4^+ in the rearing medium is varied between 3 and 560 mM (Marshall and Wood, 1990). Hemolymph Na^+ concentration is also well regulated, increasing less than 20% above the control value of 115 mM when the Na^+ concentration of rearing medium is increased from 15 to 510 mM. Potassium levels in the hemolymph vary from 13 to 36 mM when rearing medium K^+ concentration is increased from 31 to 460 mM. Marshall and Wood (1990) also report the somewhat surprising finding that the optimal pH for growth of *Lucilia* larvae is 8–9, even though much of the ammonia in the medium is then present in the non-ionic and presumably, more permeant form (NH_3). The hindgut has been proposed as the site of NH_4^+ excretion (Marshall and Wood, 1990), based on early reports of high levels of NH_4^+ in the hindgut tissues of *L. cuprina* larvae and also on the demonstration of active secretion of NH_4^+ by larvae of *Sa. bullata* (Prusch, 1974, 1976).

Another group of insects which are faced with excess ammonia are adult mosquitoes. More than 60% of the amino acids derived from the digestion of blood meal proteins by *Ae. aegypti* females is oxidized to CO_2 to provide energy and only 3% is used for egg protein production. Mosquitoes tolerate potentially toxic loads of ammonia resulting from amino acid catabolism by high rates of ammonia excretion in the feces coupled with the synthesis of glutamine and proline (Goldstrohm *et al.*, 2003; Scaraffia *et al.*, 2005). Ammonia subsequently recovered from proline can be excreted and the carbon skeleton used for synthesis of different compounds or for energy production. Ammonia excretion from mosquitoes fed 80 mM NH_4Cl in 3% sucrose increases markedly for the first day and reaches a peak of ~120 nmol animal^{-1} by 2 days, approximately sixfold greater than the level of uric acid excretion. Hemolymph glutamine and proline concentrations increase markedly within an hour when females are fed solutions containing NH_4Cl or a blood meal, whereas concentrations of aspartate, asparagine, glutamate and alanine do not change after NH_4Cl feeding. Glutamine synthetase is an important enzyme in the process of ammonia storage, since addition of an inhibitor of the enzyme added to the meal reduces hemolymph glutamine concentration and causes a corresponding increase in proline concentration. Glutamate synthase contributes to the production of glutamate for proline synthesis since azaserine, an inhibitor of glutamate synthase, results in an increase in glutamine concentration and a corresponding decrease in proline concentration after feeding. The mRNA expression and activity of several enzymes implicated in ammonia metabolism increases in the fat body after a blood meal, suggesting that it is the main tissue involved in ammonia detoxification.

5.2.3 *Uptake of ammonia*

Ammonia transport has also been studied in the caterpillar *Ma. sexta* (Weihrauch, 2006). All three sections of *Ma. sexta* midgut (anterior, median, posterior) actively absorb from the gut lumen toward the hemolymph (Weihrauch, 2006). Larvae feed on a diet which is relatively low in protein, suggesting that active ammonia absorption might be necessary for larval protein anabolism. Active ammonia uptake, which is *de facto* a net acid absorption, may also enhance midgut alkalinization. A model of ammonia uptake across the midgut epithelium suggests that NH_4^+ may enter the columnar cells via apical H^+/cation exchange, possibly by an amiloride-sensitive member of the NHE family. In addition, ammonia may enter as NH_3 across the plasma membrane, exploiting the high pH (10–11) of the lumen contents relative to that of the cells (~ pH 7). Dissociation of cytosolic NH_4^+ to H^+ and NH_3 is accompanied by diffusion of NH_3 into vesicles acidified by a bafilomycin-sensitive V-type H^+-ATPase. It is further proposed that the ammonia-loaded vesicles then move via colchicine-sensitive microtubules to the basal membrane, where they fuse with the membrane, releasing NH_4^+ into the hemolymph space (Weihrauch, 2006). In addition, a small amount of ammonia may exit from cell to hemolymph through Ba^{2+}-sensitive K^+ channels.

One consequence of ammonia uptake by the gut is that hemolymph ammonia concentration of the fifth instar larvae of *Ma. sexta* is 0.87 mM, more than twice as high as blood concentrations found in the human portal vein and about eight times as high as ammonia concentration levels in body fluids of aquatic crustaceans. Caterpillars may tolerate high ammonia levels in the hemolymph in part because high levels of a Rhesus-like ammonia transporter in the ganglia may protect neuronal tissues from ammonia toxicity. In addition, excess hemolymph ammonia may be secreted by the Malpighian tubules. Presumably ammonia is then released for excretion into the hindgut where, in contrast to midgut sections containing ~ 1 mM ammonia, very high ammonia concentrations (ca. 60 mM) are found. High levels of ammonia in the lumen of the hindgut suggest that ammonia is a major nitrogenous excretion product in *Manduca* larvae. In contrast to the midgut, the hindgut of the larvae exhibits no transepithelial active net transport, and high concentrations of ammonia in the hindgut lumen reflect entry via the Malpighian tubules and release of metabolic ammonia across the apical membrane.

Evidence for a Rhesus-like ammonia transporter (RhMS) was found with high expression levels in the hindgut, Malpighian tubules and ganglia of *Manduca* larvae but low RNA expression in midgut tissues. A full length cDNA (RhMS) has been transcribed from midgut mRNA and translation of the open reading frame indicates a 425 amino acid protein with 10 predicted transmembrane domains and more than 40% identity in amino acid sequence to known mammalian Rhesus-like ammonia transporters (RhAG, RhBG and RhCG). Low levels of mRNA expression in all midgut sections, and the fat body suggest a minor, probably regulatory role for RhMS in 'ammonia management' in these tissues (Weihrauch, 2006).

5.3 FREE AMINO ACIDS

The blood volume of insects varies greatly during changes in hydration state or in association with food intake, suggesting that the hemolymph may act as a reservoir for water. This capacity is probably a consequence of respiratory gas exchange via the tracheal system, so that changes in hemolymph volume do not impede transfer of oxygen to the tissues. In spite of changes in hemolymph volume, levels of amino acids in the hemolymph remain elevated. Whereas vertebrates have about 0.05 g% free amino acids in the blood, the amount ranges from 0.29–2.43 g% in the hemolymph of insects (Wigglesworth, 1972).

One role for hemolymph amino acids is to act as metabolic substrates for energy consuming tissues such as the flight muscles and the hindgut. Transport of amino acids has been examined in detail in the desert locust *Schistocerca americana.* Hemolymph proline and glutamate are actively secreted by the Malpighian tubules, and are subsequently reabsorbed in the rectum. The primary function of this translocation of amino acids appears to be the provision of metabolic substrates for the highly active rectal epithelial cells. However, only a small fraction of proline, alanine, serine or threonine that are actively transported across the epithelium are subsequently metabolized. Glutamate, though metabolized, is not transported across the epithelium. Surprisingly, the amount of proline transported across the rectum exceeds by a factor of 10 the amount necessary for metabolism of the rectal tissue. This extra capacity is associated with rectal water transport (Phillips *et al.*, 1986). In other words, proline may function as an osmolyte driving water reabsorption from the rectal lumen against the large osmotic differences which develop as the rectal contents are dehydrated. Neutral amino acids are also reabsorbed, probably by a common carrier system because addition of one substrate such as L-serine competitively inhibits uptake of a second, such as glycine (Balshin and Phillips, 1971).

In larvae of osmoconforming mosquitoes, proline functions as both an extracellular and intracellular osmolyte. Larvae of *Culex tarsalis* increase the hemolymph levels of both proline and the disaccharide trehalose in response to increases in environmental salinity (Patrick and Bradley, 2000a,b). In 64% sea water, hemolymph proline and trehalose are accumulated approximately 50-fold and twofold, respectively, relative to freshwater values. Proline accumulation is cued by increases in salinity, and the proximal cue for an increase in proline synthesis appears to be a rise in hemolymph Na^+. In contrast, trehalose increases in response to changes in osmolality produced either by addition of either electrolytes (sea salt) non-electrolytes such as sorbitol to the rearing medium (Patrick and Bradley, 2000b).

The amino acid tryptophan is of special significance in insects because of its role as a precursor of eye pigments. Excretion of tryptophan metabolites is a consequence of the ommochrome pathway used for synthesis of eye pigments. Eye-color mutants and their corresponding genes can be used as scorable marker systems to follow the results of genetic crossing experiments in *Drosophila* and other insects. The initial step in ommochrome synthesis involves the conversion of tryptophan to *N*-formylkynurenine by tryptophan oxygenase. In the lepidopteran, *Plodia interpunctella*, silencing of tryptophan oxygenase through the use of RNA interference (RNAi) during embryonic development results in loss of eye-color pigmentation

(Fabrick *et al.*, 2004). In the stick insect, end products of tryptophan metabolism are the ommochromes ommin and xanthommatin, which accumulate in the epidermis, and kynurenic acid, which accumulates in the faeces (Stratakis, 1979). Excretion of tryptophan metabolites accounts for up to 2% of nitrogen excretion in cockroaches, and probably represents a means of dealing with excess dietary tryptophan rather than an important component of nitrogen excretion (Mullins and Cochran, 1973a).

The most detailed studies of the role of the Malpighian tubules in excretion of eye pigments have been done on *Drosophila*. Tearle (1991) has examined tissue-specific effects of ommochrome pathway mutations in *Drosophila*, including those that alter excretion of pigment by the tubules. Dow and Davies (2003) note that the first *Drosophila* mutation described (*white*) was for eye color and that now over a hundred distinct loci are known that produce an eye-color phenotype. The *white* gene encodes an ABC transporter found in the eye and in the Malpighian tubule. Many of the eye color mutants have Malpighian tubule color phenotypes, possibly reflecting an original excretory function of the pathways that generate and transport the major classes of metabolite (ommochromes and pteridines). These pathways may have subsequently taken on specialized roles in pigment granule formation (Dow and Davies, 2003).

5.3.1 *Phloem-feeders*

Phloem-feeders such as silverleaf whiteflies (*Bemisia tabaci* Grennadius) ingest a diet that is rich in soluble carbohydrates and also contains relatively high levels of free amino acids (Buchanan *et al.*, 2000). Although plant phloem sap is generally considered to be limiting in nitrogen (Byrne and Miller, 1990), whiteflies and other phloem-feeding insects such as aphids excrete amino acids, especially non-essential amino acids such as glutamine, asparagine, glutamate and aspartate (Douglas, 1993). These four amino acids comprise the majority of the free amino acids commonly found in phloem sap (Douglas, 1993) of cotton and plants in general (Buchanan *et al.*, 2000).

The amount of amino nitrogen excreted per day exceeds by nearly twofold the total amino acid pool in the body of white-flies reared on well-fertilized cotton plants. Phloem sap of these plants thus contains supra-optimal amounts of nitrogen with regard to whitefly nutrition. However, after just a few days of feeding on low nitrogen plants, excretion of amino nitrogen essentially stops and there is an adjustment of pools of individual amino acids within the insect bodies (Crafts-Brandner, 2002). Total nitrogen content of whiteflies reared on plants with reduced nitrogen content does not decline significantly relative to whiteflies reared on plants with higher levels of nitrogen. However, the free amino acid content of whiteflies feeding on low-nitrogen plants declines $>90\%$ relative to controls. The predominant amino acids in whiteflies are glutamine (26% of the total), alanine (19%), proline (13%) and glutamate (10%), whereas the predominant amino acids in phloem sap from high nitrogen cotton plants are aspartate, glutamate and arginine, with relatively large amounts of glutamine and asparagine. The predominant free amino acids in the honeydew are asparagine (46% of total amino acid content) and glutamine (12%).

Given that the amino acid pools in whiteflies feeding on high nitrogen cotton plants are not closely related to the amino acid composition of the phloem sap, dietary amino nitrogen must be rapidly assimilated into metabolites or protein. Synthesis of at least some of the essential amino acids in aphids is dependent on endosymbiotic bacteria (Sasaki and Ishikawa, 1995; Wilkinson and Douglas, 1996; Douglas, 1998). High levels of free proline and alanine in whiteflies indicate an important role for proline metabolism in energy balance in the whitefly. The hemolymph and flight muscle tissue of several insect species contains high levels of proline that is used for fuel, via oxidation to alanine, for flight (Gäde, 1992).

It also appears that glutamine metabolism plays an important role when whiteflies adjust to changes in dietary nitrogen levels. Presumably a large and alterable pool of glutamine permits maintenance of metabolite synthesis via transamination reactions and/or formation of tricarboxylic acid cycle intermediates. Changes in the glutamine pool may be necessary to maintain levels of proline and alanine that are likely to be important metabolic fuels during flight (Crafts-Brandner, 2002).

6 Transport and excretion of divalent ions and bicarbonate

6.1 CALCIUM

Vertebrates and insects appear to use fundamentally different mechanisms to regulate the level of Ca^{2+} in their extracellular fluids. Whereas calcium homeostasis in mammals relies upon precise control of transcellular calcium absorption by the duodenum, control of excretion and not absorption is the major means of hemolymph calcium regulation in flies (*Calliphora vicina*; Taylor, 1985). When fed 12.5 mM $CaCl_2$ solution in place of water the calcium content of *Ca. vicina* rapidly increases to more than double the original content but then remains stable. However, midguts isolated from flies fed either 12.5 mM $CaCl_2$ solution or water absorb calcium at the same rate, and this rate is unaltered by treatment with the putative transport stimulants dopamine, octopamine, proctolin or 5-hydroxytryptamine. Rapid calcium absorption from the blowfly midgut appears to be an adaptation not for regulation of hemolymph calcium concentration but for removal of calcium from the lumen. This removal may facilitate other midgut functions such as phosphate absorption that would be impeded by high Ca^{2+} levels in the gut lumen as the gut contents are concentrated by water absorption. The calcium content of the Malpighian tubules is higher than that of other tissues, and rates of turnover of ^{45}Ca when flies are fed excess calcium are highest in the Malpighian tubules, indicating the tubules as the means by which hemolymph calcium concentration is regulated (Taylor, 1985).

Subsequent studies of fruit flies have shown that calcium homeostasis is accomplished through the combined effects of elimination of calcium in fluid secreted by the Malpighian tubules and the sequestration of calcium in granules, especially within the distal segment of the anterior pair of Malpighian tubules (Dube *et al.*, 2000a,b). Adult flies ingest 30 times more calcium than they retain, yet they regulate hemolymph calcium concentration and whole animal calcium content quite precisely. In response to a 6.2-fold increase in dietary calcium level, for example, calcium content of whole flies increases only 10%. Hemolymph calcium concentration ($\sim$0.5 mM) is similar in flies raised on diets differing more than sixfold in calcium content (Dube *et al.*, 2000a).

Studies with isolated tubules of *D. melanogaster* indicate that all segments of both the anterior and posterior Malpighian tubules transport Ca^{2+} at high rates and are thus capable of regulating whole animal and hemolymph calcium levels (Dube *et al.*, 2000b). Approximately 85% of calcium which enters the tubule is sequestered, and $\sim$15% is secreted in soluble form into the tubule lumen. Tubules secreting fluid at maximal rates can remove an amount of calcium equal to the whole animal calcium content in $\sim$ 9 h. The distal segments of the anterior pair of Malpighian tubules can sequester the same amount of calcium in < 2 h. The processes of secretion and sequestration are controlled independently. For example, although cAMP increases both basolateral uptake and transepithelial secretion of Ca^{2+}, the calcium channel blockers diltiazem and verapamil inhibit only basolateral uptake. Increases in basolateral membrane potential in response to reduction of bathing saline K^+ concentration enhance uptake of Ca^{2+} across the basolateral membrane, further supporting a role for Ca^{2+} uptake through channels, whereas transport from cell to lumen requires some form of Ca^{2+} pump, either an-ATPase or an electrogenic exchanger (e.g. $3Na^+$:Ca^{2+}). Ca^{2+} sequestered within the tubule is released into the lumen even when basolateral uptake is blocked. Transepithelial Ca^{2+} secretion is increased by treatments which depolarize the transepithelial potential by increasing Cl^- conductance, or which acidify the secreted fluids (bicarbonate-free saline). The latter treatment may favour dissolution of intracellular granules through changes in intracellular pH.

Mineralized granules enriched in calcium, magnesium and phosphorous are also found in the lumen of the Malpighian tubules of larvae of the face fly *Musca autumnalis* (Krueger *et al.*, 1987). These mineral stores are subsequently mobilized and used to harden the cuticle of the puparium (Krueger *et al.*, 1988). The spherical granules range from 0.2 to 10 μm in diameter and are composed of many concentric layers (Grodowitz *et al.*, 1987). Layers appear to be covered by an organic-rich material which may bind one layer to the next. Calcium, phosphorous and magnesium make up 98% of the inorganic content, which is about 85% of granule weight.

The decrease in pH from 8.08 in the lumen of the distal region of the larval tubules to 7.35 in the proximal region appears to be a major effector of granule dissolution in *Mu. autumnalis*. Release of minerals from the granules increases exponentially as the bathing medium pH decreases. The calcium and magnesium salts which are released are transported from the proximal tubule lumen to the hemolymph, and this transport may sustain or augment the rate of granule dissolution. Calcium is transported directly from tubules to the cuticle of the puparium via the hemolymph. Hemolymph calcium levels then decline as calcium is deposited in the puparial cuticle. Granule deposition, in common with granule dissolution, involves a change in pH. Cuticular pH measured with flat-surface pH electrodes increases from 7 to 8.4. This shift favours precipitation of calcium and magnesium phosphates and carbonates, which have low solubility products at alkaline pH.

Calcium is also mobilized from intracellular $CaPO_4$ concretions ('spherites') in Malpighian tubules of the house cricket *Ac. domesticus* during the early phases of diuresis (Hazelton *et al.*, 2001). When tubules are treated with the diuretic hormone second messenger cAMP, there are dramatic changes in tubule ultrastructure within 15–420 s. The cytoplasm becomes extensively vacuolated and the mitochondria move into areas adjacent the transport membranes. In addition, the $CaPO_4$ spherites become associated with both mitochondria and vesicles during this period. Although some of the spherites dissolve, many appear to be exocytosed intact into the tubule lumen (Spring and Felgenhauer, 1996). Presumably, release of spherites during diuresis in effect flushes them into the hindgut. In this way, the tubule lumen does not become obstructed, as might occur if the spherites were released into the lumen during periods of slow fluid secretion.

Blood-feeders such as *Rhodnius* are faced with a considerable excess of dietary calcium when they ingest up to 10 times their own body weight in blood, which may contain 1.5 mM calcium in the case of avian blood. Calcium accumulation by Malpighian tubule cells is correlated with the appearance of membrane bound spherical concretion bodies in the upper Malpighian tubule cells (Maddrell *et al.*, 1991). The net result of calcium deposition in the tubule is that almost none is excreted from the body. By contrast, Ca distribution between hemolymph and Malpighian tubule lumen is passive in the ant *Formica*, and the concentration of calcium in the luminal fluid is 0.9 mM (Van Kerkhove *et al.*, 1989). Calcium sequestration in *Rhodnius* tubules may be necessary to prevent elevation of calcium concentrations in the lower Malpighian tubule or hindgut during water reabsorption. High concentrations of metal ions might interfere with rectal ion and water transport. Alternatively, calcium deposits might subsequently be mobilized for production of eggs or spermataphores after ecdysis of adults.

The concretion bodies within the tubule cells of *Rhodnius* appear as concentric layered structures in electron micrographs, suggesting that they are deposited in distinct phases, perhaps after each meal. Comparison of X-ray diffraction patterns from tubular hydroxyapatite and dried tubules indicate that the concretions are amorphous, in form, not crystalline. Concentrations of Ca in the tubules exceed 21 mM, 400 mM and 1 mM when the insects are fed on low-K^+ sheep blood, rabbits and chickens, respectively. For all three diets, the hemolymph calcium concentration is 8 mM, and its activity, as determined by an ion-selective electrode, is about 2.5 mM. ^{45}Ca recently incorporated in the previous 48 h into the tubules is released at the rate of about 25% per hour when the tubules are transferred to ^{45}Ca-free saline, indicating that the process is a dynamic one. This rate declines to 13% per hour for animals allowed to incorporate ^{45}Ca for 9 days following injection of the isotope into the hemolymph. Calcium sequestration within the bodies is a continuous and metabolically dependent process, but calcium becomes less accessible to the hemolymph over time, presumably since the 'oldest' calcium is near the center of the concretions. The post-prandial increase in calcium uptake by the tubules (Maddrell *et al.*, 1991) is similar to that seen for para-aminohippuric acid and uric acid (Maddrell and Gardiner, 1975; O'Donnell *et al.*, 1983). For the latter compounds it has been suggested that an inducible transport system, stimulated by some correlate of feeding, is involved. In each case, uptake rates peak 2–4 days after a blood meal and remain elevated for about 2 weeks.

6.2 MAGNESIUM

Larvae of the salt marsh mosquito *Aedes campestris* are able to grow, moult and pupate in hyperosmotic lakes found in central British Columbia. These lakes are unusual in that they contain high levels of magnesium and that the anionic component consists almost entirely of sulfate ions (Scudder, 1969). These mosquitoes cope with osmotic water loss by drinking and absorbing the water in which they live and then excreting the excess salts (Kiceniuk and Phillips, 1974), an osmotic strategy analogous to that of marine fishes. The larvae absorb into the hemolymph virtually all the water that they ingest and thus face excess loads of sodium or magnesium and considerably larger loads of sulfate (Kiceniuk and Phillips, 1974). Malpighian tubules of the larvae secrete fluid containing 30–40 mM magnesium when bathed in solutions containing 4–20 mM Mg^{2+} (Phillips and Maddrell, 1974). Magnesium is secreted against an opposing transepithelial potential difference of 15 mV, and against concentration gradients as high as 10-fold. Transport is saturable, and the Michaelis–Menten constant for transport (K_t) is about 2.5 mM. Secretion of Mg^{2+} by the tubules permits regulation of hemolymph concentrations of 1.5–4 mM for *Ae. campestris* larvae living in and ingesting large amounts of media containing as much as 100 mM magnesium.

The highest known rates of epithelial magnesium secretion have been observed in isolated tubules of black field cricket *Te. oceanicus* (Xu and Marshall, 1999a). The small distal segment of the Malpighian tubules of the *Te. oceanicus* and the house cricket *Ac. domesticus* are characterized by very high rates of fluid secretion. Fluid secreted by the distal segment in *Te. oceanicus* is hyperosmotic and contains primarily Mg^{2+} (125 mM), Cl^- (242 mM) and Na^+ (43 mM). The concentration of magnesium in the fluid is 15 times greater than that of the bathing fluid. Surprisingly, the total volume of fluid secreted per unit time by the distal segment is constant, possibly because short distal segments are of larger diameter than long distal segments. In contrast, as the main segment increases in length the volume of fluid secreted in unit time also increases. The rate of magnesium secretion by the distal segment, 75 pmol min^{-1} mm^{-1} tubule length, is the highest known for any epithelium and is 6 times higher than the average rate of Mg^{2+} secretion by the Malpighian tubules of *Ae. campestris* (Phillips and Maddrell, 1974) and 47 times higher than the rate measured for flounder renal tubules (Beyenbach, 1990).

High rates of magnesium secretion by the distal segments of cricket tubules may play a homeostatic role by eliminating excess magnesium present in the diet. Feeding on lettuce leads to ingestion of as much as 1790 nmol of magnesium per day. All of the tubules can together secrete Mg^{2+} at the rate of 417 nmol h^{-1}, so the entire daily intake can be eliminated from the hemolymph, which contains about 431 nmol of Mg^{2+}, in about 5 h^{-1}. Some of the magnesium secreted by the distal segment appears to be reabsorbed by the main segment, raising the possibility of an interaction of the two segments in regulating hemolymph and whole animal levels of magnesium.

6.3 SULFATE

As noted above, high concentrations of sulfate are also present in the magnesium-rich lakes inhabited by larvae of the mosquito *Ae. campestris*. However, the larvae maintain sulfate concentration in the hemolymph well below that of the water. When the external concentration of sulfate is increased 30-fold from 2.5 to 73 mM, the hemolymph concentration increases only fourfold, from 1.5 to 6.6 mM. Experiments using radiolabelled sulfate have shown that the regulation of sulfate is not accomplished by precipitating the anion in the tissues. However, isolated Malpighian tubules set up in Ramsay fluid secretion assays actively transport sulfate from bath to lumen against a threefold concentration gradient. The K_t for transport is 10 mM and the maximum transport rate is 50 pmol min^{-1} tubule (Maddrell and Phillips, 1975a).

It also appears that transport of sulfate by the Malpighian tubules occurs against an opposing electrical gradient of as much as 20 mV. The latter conclusion requires confirmation because the method used for measurement of the transepithelial potential has been shown in subsequent studies to be a possible source of artefacts. Although the transepithelial potential across the large diameter tubules of *R. prolixus* can be measured by simply placing one electrode in contact with the secreted fluid droplet and the reference electrode in the saline bathing the outer surface of an isolated tubule set up in a Ramsay assay (Ianowski and O'Donnell, 2001), this method is prone to errors in tubules whose diameter is less than ~50 μm. The error results partly from shunting of current along the surface of the tubule between the bathing droplet and the secreted fluid droplet (Aneshansley *et al.*, 1988). As a result, the measurements of transepithelial potential reported in Maddrell and Phillips (1975a) may be incorrect. However, by varying the level of potassium in the bathing saline, Maddrell and Phillips (1975a) demonstrated that there is no strong relationship between the resulting changes in transepithelial potential and the secreted fluid to bath concentration ratio for sulfate. These data indicate that sulfate ions are not transported into the secreted fluid merely as a response to an electrical potential gradient favouring their entry.

Calculations based on the rate of sulfate transport by isolated tubules indicate that secretion by the Malpighian tubules is sufficient to remove all of the sulfate ingested by larvae living in waters containing less than 100 mM sulfate. At higher concentrations, sulfate ions are probably also excreted elsewhere, perhaps by the rectum or anal papillae. Larvae of another mosquito, *Aedes taeniorhynchus*, which can also live under hyperosmotic conditions, can secrete into the rectal lumen a markedly hyperosmotic fluid (Bradley, 1987). A similar mechanism in *Ae. campestris* may contribute an increasing proportion of sulfate excretion as the osmotic concentration of the external medium rises. The rectum of *Ae. campestris* has an extra posterior region which is lacking in its freshwater relative *Ae. aegypti* (Meredith and Phillips, 1973). Given the important role of the posterior rectum in secretion of hyperosmotic fluids by larvae of *Ae. taeniorhynchus,* it is quite plausible that this segment of the rectum excretes sulfate ions under hyperosmotic conditions in *Ae. campestris.*

Larvae of *Ae. taeniorhynchus* are also exposed to high levels of sulfate in salt water pools. They absorb most of the water and nearly all of the sulfate (and probably the other ions as well) from the fluid that they ingest.

Much of the water ingested by unfed larvae is to replace that lost by osmosis to the hyperosmotic medium. Sulfate absorption by the midgut does not proceed against even a small concentration difference. Since most epithelia are relatively impermeable to ions as large a sulfate, these findings indicate that its uptake across the midgut epithelium probably involves some form of facilitated diffusion.

The sulfate transport system in tubules from *Ae. taeniorhynchus*, as in tubules of *Ae. campestris*, is characterized by high capacity and low affinity, transport being half saturated in the range 5–10 mM. However, tubules from insects reared in sulfate-free water have only a weak ability to transport sulfate ions; indeed the fluid they secrete is scarcely if ever found to contain sulfate ions at a higher concentration than in the bathing medium. In contrast, Malpighian tubules from *Ae. taeniorhynchus* reared in sulfate-containing waters develop an ability to transport sulfate at a rate which varies directly with the sulfate content of the environment (Maddrell and Phillips, 1978). Malpighian tubules from larvae reared in sulfate-enriched sea water containing 89 mM sulfate transport sulfate ions at rates twice those from insects reared in normal sea water and the fluid secreted contains sulfate ions at concentrations several times higher than in the bathing medium. For example, tubules bathed in fluid containing 3.3 mM sulfate secreted fluid containing an average of 18.5 mM sulfate and tubules bathed in fluid containing 30 mM sulfate secreted fluid containing more than 130 mM sulfate. The transport ability is not retained in adults which developed from larvae reared in sulfate-rich waters. The induction of transport appears to involve a change in capacity with no change in affinity and the half-time for induction of this transport ability is approximately 16 h. The sulfate-transporting ability induced in the Malpighian tubules is more than sufficient to match the rate at which sulfate ions are taken up across the gut. Induction may involve one or more of the following mechanisms: synthesis of more pump molecules, the incorporation of more pumps into the cell membranes, or the activation of inactive pumps already in the membranes (Maddrell and Phillips, 1978).

The rate of sulfate transport by the isolated Malpighian tubules depends on the rate of fluid secretion. Fluid secretion rates by the Malpighian tubules can be increased up to sixfold above the basal level by addition of extracts of the brain and of the thoracic ganglia. Sulfate ions are removed more slowly from the bathing medium at the lower rates of fluid secretion. This relationship is very similar to that seen in the transport of organic anions and cations by Malpighian tubules of other insects (Maddrell *et al.*, 1974; Maddrell and Gardiner, 1976). The passive permeability of the tubule wall reduces the effectiveness of transport at the lower rates of fluid secretion through passive diffusive backflux of the transported molecules from the lumen to the bath, as described in Section 8.3.

Malpighian tubules of the blowfly *Calliphora erythrocephala* also secrete fluid containing levels of sulfate above those in the bathing medium over the range of bathing medium sulfate concentrations from 1 to 50 mM (Knowles, 1975b). Earlier studies (cited by Knowles, 1975b) show that tubules *Calliphora* as well as those of the hemipteran *Rhodnius* and the stick insect *Ca. morosus* all cease fluid secretion when sulfate is the only anion present in the bathing fluids. However, lower concentrations of sulfate do not suppress fluid secretion by tubules of *Calliphora* or *Carausius*. For *Calliphora* tubules, the highest ratio of sulfate in the secreted fluid to that in the bathing medium (SF/M) is 4.8 and is found when the bath contains 6 mM sulfate. Intact flies also excrete sulfate in the urine.

The mechanism of sulfate transport by *Calliphora* tubules is unclear. The measured transepithelial potential is not sufficiently positive to account for the observed concentrations of sulfate in all experiments, and the 40-fold increase in sulfate transport between 0.5 and 5 mM sulfate in the bathing saline is not associated with a sufficiently large change in transepithelial potential to explain the increase. However, the method of transepithelial measurement used in this and other early studies has subsequently been shown to result in erroneous values in small diameter tubules such as those of blowflies and mosquitoes, as discussed above.

The function of sulfate transport in insects not normally exposed to high levels of this anion is unknown. However it has been pointed out by Knowles (1975b) that sulfate secretion may be linked to the metabolism of hormones. Malpighian tubules of *Ca. erythrocephala* conjugate β-ecdysone with sulfate and metabolites of insect juvenile hormone can combine with sulfate to form conjugates which are then excreted.

6.4 BICARBONATE

6.4.1 *Transport of bicarbonate by Malpighian tubules*

Bicarbonate transport by insect epithelia has been examined primarily in species adapted to alkaline lakes. For the freshwater species of water boatmen, *Cenocorixa blaisdelli* (Corixidae, Hemiptera) it appears that extrarenal pathways contribute to hemolymph ion and CO_2 regulation when the animals are acclimated to high pH ($NaHCO_3$, Na_2CO_3) water typical of athalassohaline lakes (i.e. those not of marine origin but from evaporation of fresh water) found in western Canada (Cooper *et al.*, 1987). Although hemolymph ion concentrations in alkaline-acclimated insects do not differ from those of insects reared in fresh water, urinary loss of ions and CO_2 cannot fully compensate for increased intake of sodium and CO_2. Chloride cells are present on the bodies (Komnick and Schmitz, 1977) of corixids and on the labium of the alkaline lake species *Cenocorixa bifida* (Jarial *et al.*, 1969), raising the possibility that increases in hemolymph chloride concentration under alkaline conditions reflect extrusion of basic equivalents by Cl^-/HCO_3^- exchange (Cooper *et al.*, 1987). Regulation of hemolymph Na^+, however, is more problematic. The possibility of extrarenal Na^+/H^+ exchange seems unlikely since, the elevated pH and high sodium content of the surrounding water would make Na^+ uptake more likely than Na^+ loss. It is worth noting that changes in osmotic water influx may also be important in ion-regulation by corixids. The water permeability of *Ce. bifida* increases when ambient salinity increases, so corixids may be able to obtain some water without an ion load if the cuticle acts as an ion filter during osmotic influx of water (Cannings, 1981).

A role for the Malpighian tubules in secretion of bicarbonate has been suggested by studies of fresh water and alkaline lake species of corixids. Four distinct segments are present in Malpighian tubules of these species. Segment 1 is connected to the hindgut and appears to act as a passive conduit. In response to cAMP, segment II increases the pH of the luminal fluid, from 7.0 to 8.1 in the fresh water species *Ce. blaisdelli* (Cooper *et al.*, 1989), and from 7.0 to 8.6 in the alkaline lake species *Ce. bifida* (Szibbo and Scudder, 1979). The alkalinity increase is correlated with increased total CO_2 content of luminal fluid, suggesting that bicarbonate is secreted by segment II. The carbonic anhydrase inhibitor acetazolamide blocks the pH change, and does so in the absence of exogenous bicarbonate, suggesting that bicarbonate in the lumen is derived from metabolically produced CO_2. cAMP may activate carbonic anhydrase, therefore, or a bicarbonate conductance. The positive transepithelial potential would favour anion movements into the lumen, although a cell-to-lumen concentration gradient of bicarbonate is presumably required as well to explain an elevation of pH by one unit or more.

Transport and storage of bicarbonate and carbonate play important roles in the biology of shore flies (Ephydridae, Diptera). These insects are well known for their tolerance of extremes in temperature, salinity and pH which occur in the larval habitats such as hot springs, hypersaline desert lakes and tidal splash pools. *Ephydra hyans* inhabits alkaline salt lakes in western North America, and is capable of hyperosmotic and hypoosmotic regulation of hemolymph osmolality (at around $300\,mOsm\,kg^{-1}$) in alkaline lake water, but less capable of regulation in non-alkaline sea water/sodium chloride solutions (Herbst *et al.*, 1988). Although hemolymph osmolytes consist primarily of sodium and chloride, carbonate and bicarbonate are major components of the species habitat. The Malpighian tubules on one side of the gut appear to excrete and store an insoluble salt of CO_3^{2-} and HCO_3^-. Although the morphology of the anterior pair of Malpighian tubules is typical of most insects, the posterior pair of tubules is modified into an enlarged gland containing spherical concretions of nearly pure calcium carbonate (Herbst and Bradley, 1989). These lime gland tubules accumulate calcium more rapidly than normal tubules.

It appears that carbonate and bicarbonate are regulated by means of chemical precipitation of these compounds with calcium in the lumen of the lime gland tubules Herbst *et al.* (1988). This storage excretion role of the Malpighian tubules contrasts with the structural modification of ileum and rectum that appear to be involved in ionoregulation by other ephydrids (Marshall and Wright, 1974).

Lime gland mass increases with the proportions of HCO_3^-/CO_3^{2-} in the surrounding waters, and also with increased osmolality due to NaCl. It has been suggested that the lime gland serves not only to regulate carbonates, but to provide negative buoyancy (Herbst and Bradley, 1989). Larvae will tend to float and be exposed to higher mortality in wave-washed shallows of their lakeshore habitat in the absence of a lime gland 'weight belt'.

6.4.2 *Transport of bicarbonate by the rectal salt gland of mosquito larvae*

Larvae of the mosquito *Aedes dorsalis* inhabit alkaline (pH 10.5) lakes rich in bicarbonate (250 mM) and carbonate (100 mM). Although larvae ingest the lake water at rates up to 130% body weight per day, they maintain hemolymph pH between 7.55 and 7.70 and hemolymph bicarbonate concentration between 8 and 18.5 mM (Strange *et al.*, 1982). Regulation is achieved through active secretion of HCO_3^- and CO_3^{2-} by the rectal salt gland. The rectum is divided into two structurally distinct regions termed the anterior and posterior segments. Hyperosmotic secretion occurs in the posterior segment, whereas chloride uptake by exchange with HCO_3^-, along with organic solute reabsorption, occurs in the anterior segment (Bradley, 1987). Bicarbonate and carbonate are excreted against both chemical and electrical gradients. The lumen:hemolymph chemical gradient is 21:1 for HCO_3^- and 241:1 for CO_3^{2-}, and the rectal transepithelial potential is 14–25 mV, lumen negative. The net electrochemical potential is 69 mV for HCO_3^- and 76 mV for CO_3^{2-}, indicating active transport of both ions into the lumen.

Salt glands isolated and microperfused *in vitro* maintain transepithelial potentials as high as −50 mV, lumen negative, and secrete CO_2 in the absence of Na^+ or K^+ in lumen or serosal bathing salines. Bicarbonate secretion is inhibited by removal of chloride or addition of acetazolamide or DIDS, and net chloride reabsorption equals the rate of total CO_2 secretion, suggesting a 1:1 exchange of luminal Cl^- for serosal HCO_3^- (Strange and Phillips, 1984).

Both CO_2 secretion and Cl^- reabsorption occur primarily in the anterior rectal gland, whereas the posterior segment is responsible for 75% of total hyperosmotic NaCl secretion, with the anterior segment contributing the remainder (Strange *et al.*, 1984). It seems likely that both segments secrete NaCl rich fluids *in vivo*, and that the composition of the luminal fluid is subsequently modified by ion exchange and reabsorptive processes.

Fluid movements *in vivo* are not unidirectional or continuous, since the salt gland empties only after becoming filled with secretion and faecal material. Fluid secreted by the posterior segment, therefore, can come in contact with, and be modified by, the anterior segment.

A tentative model of ion transport by the gland was developed based on data from recordings with intracellular voltage-, pH- and Cl^--selective microelectrodes, coupled with calculations of bicarbonate activity by the Henderson–Hasselbalch equation (Strange and Phillips, 1985). Bicarbonate crosses the basolateral membrane into the anterior rectal cell by an active mechanism against an electrochemical gradient of 77 mV and exits the cell at the apical membrane down a favourable electrochemical gradient of 28 mV (Fig. 10). The model proposes that chloride crosses the apical membrane into the cell by passive, electrodiffusive movement through a chloride-selective channel, and bicarbonate moves from cell to lumen by a passive or active electrogenic mechanism. Changes in serosal Cl^- concentration have no effect on basolateral membrane potential, indicating an electrically silent mechanism for Cl^- movement from cell to hemolymph. This step is DIDS-sensitive, raising the possibility of direct coupling of chloride and bicarbonate movements through Cl^-/HCO_3^- exchange across the basolateral membrane (Strange and Phillips, 1985).

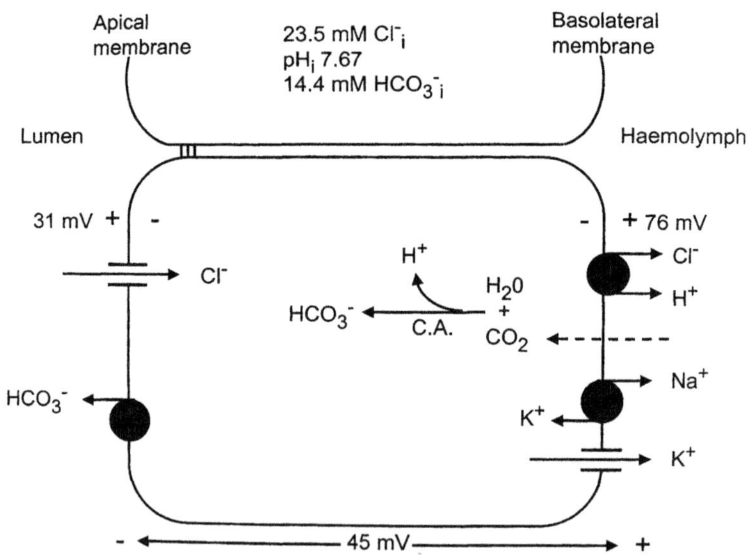

FIG. 10 Tentative cellular model of HCO_3^- and Cl^- entry and exit steps in anterior rectal salt gland cells of the larva of the mosquito *Ae. dorsalis* (c.a. = carbonic anhydrase). Note that the actual species (i.e. HCO_3^-, H^+ or OH^-) involved in transepithelial HCO_3^- secretion are not known with certainty. Redrawn from Strange and Phillips (1985, Fig. 12). Reprinted with permission from the American Physiological Society.

7 Excretion and sequestration of toxic metals

7.1 SITES AND MECHANISMS OF CADMIUM ACCUMULATION AND TRANSPORT

High levels of metals are accumulated by the midgut epithelium in many insects (e.g. Postma *et al.*, 1996), in part in metal-containing mineral concretions. In ants, for example, measurements of metal levels (Pb, Cd, Cu, Zn, Fe, Mn) in different tissues of workers of *Formica pratensis* Retzius, *F. polyctena* (Forster) and *Camponotus ligniperda* (Latreille) collected from a metal-polluted site reveal that highest levels are found in the midgut, followed by the Malpighian tubules and the hindgut (Rabitsch, 1997). Levels of the metals in the tissues are correlated with the extent of metal pollution of the collection sites. The midgut appears to act as an effective barrier for many of the minerals, thereby limiting the extent to which toxic metals cross into the hemolymph, although the Malpighian tubule cells also contain concretions which may act as a form of storage excretion. A switch from accumulation to elimination presumably occurs in long-lived species when the storage capacity is exceeded.

Microautoradiography of whole larvae of *Chironomus riparius* exposed to cadmium in water or in water and sediment has shown that the alimentary canal is the primary site of cadmium accumulation (Craig *et al.*, 1998). The digestive tract of *Ch. riparius* larvae has four structural divisions: the oesophagus (stomadeum), the anterior midgut, posterior midgut and the hindgut (proctadeum). Cadmium accumulates only in the anterior portion of the posterior midgut, a section which is rich in mitochondria, particularly within the long apical microvilli and alkaline phosphatase. These characteristics are consistent with a role in absorption. Accumulation of cadmium by the posterior midgut is related presumably to the function of the midgut as the site of absorption of calcium, and cadmium uptake probably occurs via calcium channels (Craig *et al.*, 1998). Uptake of Cd^{2+} via Ca^{2+} channels is to be expected on the basis of similarity in ionic radii (0.097 and 0.099 nm for Cd^{2+} and Ca^{2+}, respectively). Consistent with a role for Ca^{2+} channels, Cd^{2+} accumulation can be reduced as much as 88% by increasing the Ca^{2+} concentrations of the water from 0.1 to 10 mM (Craig *et al.*, 1999). There is also a concentration-dependent decrease in Cd^{2+} accumulation in response to the Ca^{2+} channel blockers lanthanum (73% at 10 μM and 92% at 100 μM) and verapamil (59% at 100 μM and 85% at 300 μM; Craig *et al.*, 1999). A role for Ca^{2+} channels in Cd^{2+} uptake has also been proposed from studies of the *Aedes albopictus* C6/36 cell line. Uptake is saturable and the value of the Michaelis–Menten constant (K_t) is 1.7 μM. Uptake is unaffected by pre-treatment with dinitrophenol but is reduced by verapamil, suggesting that the metal is taken up into the cells by mediated transport not requiring metabolic energy (Braeckman *et al.*, 1999).

The latter study also indicated that production of proteins which are likely members of the heat shock protein (HSP) family is induced by Cd^{2+} exposure. HSPs, metallothioneins and a family of proteins in invertebrates known as cadmium-binding proteins (CBPs) are all known to bind cadmium.

The Malpighian tubules were included with the gut in the analyses of the larvae of *Ch. riparius* by Postma *et al.* (1996), and separate values for the levels of cadmium in each tissue, therefore, are not available for this species. Seidman *et al.* (1986b) note little accumulation of radiolabelled cadmium in the anterior midgut, exoskeleton or Malpighian tubules, although the loss of more than 30% of the isotope during fixation of the tissue reduced the sensitivity of their technique. However, in the silkworm (*B. mori*) reared on an artificial diet containing cadmium at concentrations of 5 and 80 $\mu g\,g^{-1}$ wet diet, cadmium accumulated in the alimentary canal and the Malpighian tubules at concentrations of 1100 and 470 $\mu g\,g^{-1}$ wet weight, respectively. The finding of high levels of cadmium in the Malpighian tubules indicates that cadmium crosses the gut epithelium into the hemolymph, and that it is then transported into the Malpighian tubules (Suzuki *et al.*, 1984).

Metal-adapted and non-adapted populations of *Ch. riparius* differ in rates of storage and excretion of cadmium (Postma *et al.*, 1996). The adapted populations both accumulate more cadmium from the water and show higher rates of cadmium elimination when transferred to cadmium-free water. Cadmium may be initially sequestered by metal-binding proteins but eventually it is stored in the form of membrane bound concretions that have been observed in the posterior midgut of *Chironomus* (Seidman *et al.*, 1986a; Craig *et al.*, 1998). Cadmium-containing granules in the midgut may be expelled into the lumen by exocytosis or degeneration of whole cells. Degeneration of posterior midgut cells is seen in larvae exposed to high concentrations of cadmium (Seidman *et al.*, 1986a). The effects of crossing adapted and non-adapted chironomid populations on larval growth and EC_{50} values during cadmium exposure has been examined by Groenendijk *et al.* (2002). A rapid loss of metal adaptation in the first generation hybrid offspring is evident in crossings of the adapted and non-adapted populations. Reciprocal crossings rule out any maternal effects and show a major genetic component for metal tolerance in exposed chironomid populations.

It is important to point out that the high tolerance of *Chironomus* larvae to cadmium exposure is not explicable solely by induction of cadmium-binding proteins. Reports that exposure to metals such as copper and cadmium induces metal-binding proteins in species such as the caterpillar *Galleria mellonella*, the housefly *Musca domestica*, the fruit fly *D. melanogaster* and several species of chironomids have been discussed by Postma *et al.* (1996). Duplication of the metallothionein gene may be an adaptation to heavy metals in *D. melanogaster*. In Malpighian tubules of the latter species, the importance of metallothioneins can be deduced from the finding that metallothionein A is the most abundant transcript of any gene in the Malpighian tubule (Wang *et al.*, 2004). In *Chironomus*, however, cadmium uptake is very rapid, and appears to be non-specifically absorbed onto high molecular weight proteins from which it is rapidly released when larvae are transferred to cadmium-free water. Although a low molecular weight cadmium-binding protein is induced by exposure to cadmium in the water, the induction is very slow relative to the rate of cadmium uptake (Yamamura *et al.*, 1983). The cadmium-binding protein contains little zinc or copper, in contrast to vertebrate metallothioneins. Displacement of cadmium by copper, the prevalence of cysteinyl residues and absence of aromatic amino acids in the protein are common to metallothioneins in other species. The high tolerance of *Chironomus* larvae cannot be explained by the induction of these cadmium-binding proteins because cadmium is mostly not bound to them during the period of rapid cadmium uptake (Yamamura *et al.*, 1983). Nor is complexation with proteins the sole mechanism of cadmium detoxification in *Sarcophaga peregrine*. In this insect most of the cadmium accumulated from the diet is retained through the pupal stage but is excreted by the adult during the first day after emergence. Levels of a cadmium-binding protein gradually decrease after pupation, so that although most of the cadmium accumulated in the larvae is retained in the pupae, only limited amounts of it are bound to protein. Protection of pupal tissues and those of the newly emerged adult from the toxic effects of cadmium therefore requires a mechanism other than complexation with proteins such as metallothioneins (Yamamura *et al.*, 1983).

Results of a study on predatory beetles which had been fed mealworm larvae containing Cd, Zn, Hg or methyl mercury (Me-Hg) also indicate that complexation in the midgut cannot be the sole mechanism of detoxification (Lindqvist *et al.*, 1995). It appears that loss of the metals reflects faecal excretion of metals that were not absorbed from the food plus the discharge of midgut epithelial cells which had taken up each metal. Levels of cadmium and mercury are reduced to less than 1% of the initial quantity present within 2 weeks. By contrast, 60% of Me-Hg and 20% of Zn are still present in the beetle tissues 30 d after ingestion of metal-loaded food. Slower rates of discharge of Me-Hg reflect the more hydrophobic nature of the compound and consequent bioaccumulation. Autoradiography of slices of the beetles 2–3 weeks after ingestion of cadmium show that the metal is located primarily in the midgut but also in the integument. Some cadmium must therefore cross the gut epithelium into the hemolymph for subsequent deposition in the integument. In *Drosophila* larvae which have ingested cadmium, most of the metal is found in the digestive tract, with smaller quantities present in the Malpighian tubules, hemolymph, fat body and integument (Maroni and Watson, 1985). The latter findings indicate passage of cadmium across the midgut epithelium in *Drosophila* as well. The significance of this passage is that cadmium detoxification cannot be explained solely through processes within the midgut, and a role for sequestration and or secretion within the Malpighian tubules and/or hindgut seems plausible.

Whereas terrestrial insects may face metal loading through the diet, aquatic species may also take up metals across the external organs used for ionoregulation. The number of ionoregulatory cells (chloride cells) is related to uptake rates for dissolved $Cd^{2+}$and Zn^{2+} (Buchwalter and Luoma, 2005). Dissolved Cd^{2+} uptake rates are determined by the numbers of Cd^{2+} transporters rather than by the affinities of those transporters for Cd^{2+}, and Cd^{2+} uptake is not related to body size or gill size.

7.2 REGULATED STORAGE OF ZINC AND COPPER IN *DROSOPHILA*

The presence of zinc, copper and iron in storage vesicles within the Malpighian tubules (Sohal *et al.*, 1977) lead to the hypothesis that accumulation of these metals was a detoxification mechanism. The gut may also be implicated in detoxification. For example, copper cells found in the *Drosophila* midgut are named for their capacity to accumulate the metal and are also of interest because they show structural similarities to the acid-producing gastric parietal cells of the mammalian stomach (Dubreuil, 2004). Maddrell (1977) suggested that metals would be reabsorbed into the hemolymph if they were excreted from the Malpighian tubule into the gut. Metal ions are small relative to organic toxins and would therefore diffuse more readily through the cells of the hindgut into the hemolymph and would thus simply be returned to the Malpighian tubules. This futile cycling would be particularly problematic if a substantial proportion of the metals passing through the gut were absorbed, and the storage excretion of metals would thus minimize energy expenditure and protect sensitive tissues from exposure to toxic levels of divalent cations.

The detoxification hypothesis has been challenged by Schofield *et al.* (1997), who propose that accumulation of Zn and Cu in tissues of the fruit fly *Drosophila* represents regulated storage rather than deposit excretion. Several pieces of evidence support this view. Measurement of elemental contents of whole flies and organs by a penetrative ion microprobe technique shows that zinc is accumulated primarily in the main segments of both the anterior and posterior Malpighian tubules at concentrations up to 2.8% of tissue dry weight (Schofield *et al.*, 1997). By contrast, iron and copper are distributed throughout the body. Accumulation of a specific divalent metal is considered to indicate regulation of biological availability rather than deposit excretion. The most critical challenges to the detoxification hypothesis were the findings that less than 1% of copper and zinc are absorbed from the food and that there is no proportionality between the quantities of zinc and copper consumed and the quantities accumulated.

Regulated storage of zinc rather than deposit excretion is consistent with the requirement of zinc in several biological processes. In contrast to copper and cadmium, zinc does not bind to metallothioneins nor does it induce their synthesis (Maroni and Watson, 1985). Although high dosages of zinc are toxic, zinc is a component of at least 300 enzymes and is also chelated by zinc finger proteins which are implicated in gene regulation. Zinc is also incorporated into the arthropod cuticle during moulting and is accumulated at concentrations as high as 25% in the mouthparts and claws as a means of increasing hardness.

7.3 DETOXIFICATION AND EXCRETION OF IRON AND HEME

Multiple tissues are involved in maintaining appropriate levels of iron within insects (Locke and Nichol, 1992). The midgut plays roles in uptake of iron from the diet, storage and also excretion. The hemolymph plays a role in transport and storage, and the Malpighian tubules are also implicated in uptake from the hemolymph, storage and excretion. Lastly, the pericardial cells are important for uptake and heme recycling and the fat body plays a role in intermittent storage of iron. Within cells, diverticula of the rough endoplasmic reticulum contain concretions of calcium salts along with other metals such as zinc and iron.

Blood-feeders, such as mosquitoes and the hemipteran *R. prolixus*, face a massive release of heme during hydrolysis of host hemoglobin in the digestive tract. Subsequent degradation of heme by heme oxygenase (HO) leads to production of iron, carbon monoxide and biliverdin. Both iron and heme lead to formation of reactive oxygen species that can damage a variety of biological molecules. Although biliverdin does have antioxidant properties, it must be excreted to avoid cytotoxicity presumably related to the effects of high levels of hydrophobic molecules on membrane structure and function as well as the effects of bile pigments on DNA and protein synthesis.

Studies by Paiva-Silva *et al.* (2006) have identified a unique and complex heme-degradation pathway, with four distinct intermediates, in *R. prolixus*. In mammals the heme ring is cleaved by HO at the alpha methene bridge to form biliverdin which is subsequently converted to bilirubin by biliverdin reductase. In *Rhodnius*, heme is first modified by addition of two cysteinylglycine residues before cleavage of the porphyrin ring, followed by trimming of the dipeptides. The end product of heme detoxification in *Rhodnius* is a dicysteinyl-biliverdin IXγ. The addition of hydrophilic residues to heme molecules may be a way to facilitate the excretion of large amounts of biliverdin, a relatively hydrophobic molecule, produced after blood digestion (Paiva-Silva *et al.*, 2006). Similarly, in mammals, bilirubin is excreted only after conjugation to glucuronic acid, and transport by members of the multidrug resistance-associated protein (MRP) family and the OATP family of anion transporters has been implicated.

Although both heme and iron cause oxidative stress, they promote lipid peroxidation by different mechanisms. Iron-induced oxidative stress results from generation of hydroxyl radicals ($OH^{\bullet}$) by the Fenton reaction:

$$Fe^{2+} + H_2O_2 \rightarrow Fe^{3+} + OH^- + OH^{\bullet}$$

Hydroxyl radicals can then initiate lipid peroxidation chains by removing electrons from other molecules such as unsaturated fatty acids, thereby generating an alkyl radical ($R^{\bullet}$). In contrast, heme-induced formation of radical species results from conversion of less reactive organic hydroperoxides (ROOH) into highly reactive alkoxyl ($RO^{\bullet}$) and peroxyl ($ROO^{\bullet}$) radicals (Graca-Souza *et al.*, 2006).

An important antioxidant role is played by urate in the hemolymph of the blood feeder *R. prolixus*. Although the Malpighian tubules of *Rhodnius* secrete urate at high rates, the rates of urate synthesis and excretion are balanced so that the hemolymph contains urate concentrations as high as 5 mM (Souza *et al.*, 1997). Urate synthesis by the fat body *in vitro* is stimulated by the presence of hemin (the Fe^{3+} oxidation product of heme) through a mechanism involving protein kinase C activation. Hemolymph urate concentration increases *in vivo* in response to blood feeding or injection of hemin into the hemocoel and also when animals are exposed to hyperoxic conditions (70% O_2). Preventing urate synthesis *in vivo* by addition of the xanthine oxidase inhibitor allopurinol results in increased levels of free radicals in the hemolymph. Urate synthesis thus appears to be an important adaptation to the oxidative stress produced by hemin degradation during digestion of the blood meal. Additional protection is provided by the *Rhodnius* heme binding protein (RHBP).

In mosquitoes, protection against the oxidative challenge presented by excess iron derived from digestion of the blood meal in mosquitoes is provided by the iron-binding protein ferritin (Geiser *et al.*, 2003) and also be excretion (Zhou *et al.*, 2007). The primary site of expression of the ferritin in the yellow fever mosquito, *Ae. aegypti*, is the midgut and expression of both the ferritin heavy-chain homolog (HCH) and the light-chain homolog (LCH) are induced by blood-feeding. Messenger RNA levels for both LCH- and HCH-genes increase with iron, hydrogen peroxide and hemin treatment, and the temporal pattern of gene expression of the genes is very similar. Ferritin is also present in the peripheral tissues of blood-feeders to provide further protection against the effects of iron-induced oxidative stress. It appears that iron required for short term nutrient needs is retained in cytoplasmic ferritin, whereas excess iron is stored inside ferritin that is subsequently secreted from the cells into hemolymph. Secreted ferritin may thus be a primary iron storage protein in mosquitoes and the hemolymph may be the primary tissue involved in iron storage. Ferritin secretion may thus provide a mechanism by which mosquito cells protect against iron overload and reduce the potential for iron-mediated oxidative stress inside cells (Geiser *et al.*, 2006).

It also appears that excretion plays a major role in preventing iron toxicity after the blood meal. The blood meal provides high levels of iron primarily as hemoglobin in erythrocytes, and to a small extent as ferric-transferrin. A quantitative evaluation of the fate of blood-meal iron during the first gonotrophic cycle in *Ae. aegypti* has been provided by Zhou *et al.* (2007). Most of the heme iron provided by the blood meal is excreted, whereas iron provided as ferric-transferrin is highly absorbed and transferred to the ovaries and eggs. Because iron provided from heme in blood is in much greater quantity than that provided by transferrin, the majority of iron absorbed from the blood meal comes from heme. Of the iron supplied in the form of heme, 87% of the ingested total is excreted, 7% accumulates in the eggs and 6% is stored in other tissues. Of the iron supplied in a blood meal, approximately 7% appears in the eggs and of this iron 98% is from hemoglobin and 2% from ferric-transferrin.

A role for the Malpighian tubules in iron metabolism has been suggested on the basis of studies of the mosquito *Anopheles albimanus*. Martinez-Barnetche *et al.* (2007) cloned a homolog (AnaNRAMP) of the *d*ivalent *m*etal *t*ransporter 1/*n*atural *r*esistance-*a*ssociated *m*acrophage *p*rotein (NRAMP) family. This group of proton-coupled divalent cation transporters is widely distributed in prokaryotes and in eukaryotes and is involved in Fe and Mn transport. AnaNRAMP mRNA induces iron incorporation when injected into *Xenopus* oocytes and Western blots show expression of AnaNRAMP in head, midgut and at high levels in Malpighian tubules of unfed female mosquito. AnaNRAMP levels in the midgut decline after blood-feeding, but increase in the Malpighian tubules. AnaNRAMP is localized in the membrane of intra-cellular concretions or spherites in the principal cells of the Malpighian tubule. Localization of AnaNRAMP in concretions suggests that divalent cations such as Fe^{2+} and Mn^{2+} may also be deposited there for storage. AnaNRAMP may thus function as a divalent cation efflux pump transporting iron from the vesicle to the cytosol, were it can be stored by cytosolic ferritin. At present it is not known if AnaNRAMP-mediated cation transport is involved in iron retention/storage or iron excretion (Martinez-Barnetche *et al.*, 2007).

In honeybees iron is obtained from pollen in the diet, and the rate of iron accumulation is directly related to the levels of iron in the diet. Honey bees store iron both as granules in the midgut and in the trophocytes of the fat body. However, the iron content of the fat body reaches a maximum level regardless of the amount of iron available for ingestion, and is reached at the commencement of foraging behavior, suggesting a possible role in orientation. The honey bee iron granules are larger than other forms of iron granules (0.32 μm diameter), contain significant amounts of calcium and phosphorous, and have an amorphous electron structure when examined by X-ray diffraction. Given that magnetite is absent or only a minor component (<0.33%) and that the trophocytes are not innervated, a direct role for the granules in magnetic field detection seems unlikely. However, the granules may be paramagnetic and affect local intracellular enzymatic activity in response to fluctuations in external magnetic fields (Kuterbach and Walcott, 1986a,b; Kuterbach *et al.*, 1982). Alternatively, the granules may be involved in iron homeostasis. Many of the enzymes required for energy production during flight, for example, contain iron (e.g. cytochrome c oxidase, NADPH oxidase).

8 Transport of organic cations and organic anions

Renal tissues of vertebrates and invertebrates transport a wide range of organic cations (OCs), most with a quaternary nitrogen, a hydrophobic region and a positive charge at physiological pH. They include numerous exogenous compounds and metabolites, many of which are toxic. Transport and elimination of OCs is therefore essential for the maintenance of homeostasis. Multiple transporters are involved, in part reflecting the need to transport molecules from two broad structural classes. Type I organic cations are relatively small (< 400 Da) hydrophilic monovalent compounds and include both endogenous molecules such as choline or *N*-methylnicotinamide (NMN) and drugs such as tetraethylammonium (TEA). Type II organic cations are larger amphiphilic compounds that contain a positively charged group situated close to or within a large aromatic ring structure (Wright and Dantzler, 2004). Examples of type II organic cations are nicotine and quinidine. Organic cation transport by renal organs commonly involves carrier-mediated potential-driven uptake at the basolateral membrane, intracellular sequestration that reduces the free concentration of the cation, and luminal exit by P-gps or through organic cation-proton exchange (Pritchard and Miller, 1993). Studies of organic cation transport often utilize radiolabelled compounds such as the prototypical organic cations TEA or NMN. Organic cation transport can also be quantified through the use of fluorescent compounds such as [2-(4-nitro-2,1,3-benzoxadiazol-7-yl) aminoethyl]-trimethylammonium (NBD-TMA) (Bednarczyk *et al.*, 2000) or quinidine (Leader and O'Donnell, 2005).

8.1 TRANSPORT OF TYPE I ORGANIC CATIONS

Early qualitative studies indicated the accumulation of basic (cationic) dyes such as methyl green and methylene blue by the proximal Malpighian tubules of *Ma. sexta* (Nijhout, 1975). The first quantitative estimates of excretion of type I organic cations such as TEA or NMN in insects were presented for *D. melanogaster* (Rheault and O'Donnell, 2004). Electrophysiological techniques for assessing the fluxes of tetraalkylammonium compounds, especially the prototypical organic cation TEA, were applied to analysis of organic cation transport by isolated guts and Malpighian tubules. The techniques exploit the high selectivity of the cation exchanger potassium tetra(*p*-chlorophenyl)borate (Corning 477317) for quaternary ammonium compounds such as TEA and tetramethylammonium (TMA). Although originally used for measurement of K^+ activity, the selectivity of microelectrodes based on this exchanger for TEA and TMA exceeds that for K^+ by factors of 10^7 and $10^{2.7}$, respectively.

TEA-selective microelectrodes are used to measure TEA transport in two ways, and related microelectrode methods can be applied to measuring the transport of other ions (Na^+, K^+) and also organic compounds such as salicylate (Section 8.3) by Malpighian tubules and other tissues. Figs. 11–14 are therefore presented both with a view to explaining specific features of TEA transport and with a view to showing the methodologies for quantifying ion fluxes through the use of ion-selective microelectrodes. Firstly, the concentration of TEA is measured in fluid droplets secreted by isolated Malpighian tubules of *Drosophila* set up in a Ramsay secretion assay (Fig. 11). Flux across the entire tubule is calculated from the product of secretion rate and secreted droplet TEA concentration. Secondly, TEA-selective microelectrodes based on Corning 477317 ion exchanger are also used with the scanning ion electrode technique (SIET) to assess spatial and temporal variations in TEA flux in different regions of the Malpighian tubules and gut. A TEA-selective microelectrode is moved between two positions within the unstirred layer next to the surface of a cell, and the measured difference in TEA concentration between the two positions is converted into a corresponding TEA flux using the Fick equation (Fig. 12).

Measurements based on the Ramsay assay and the SIET measurements have revealed TEA transport by the Malpighian tubules, ureter and posterior midgut (Figs. 13, 14). TEA transport by the posterior midgut may serve to minimize absorption of potentially toxic organic cations from the gut lumen into the hemolymph, whereas transport by the tubules presumably serves to clear the hemolymph of toxins. Given that the transepithelial potential across the main (secretory) segment of the tubule is 30–80 mV lumen-positive (O'Donnell *et al.*, 1996) and that the lumen concentration of TEA is ~12-fold above that in the bath for main

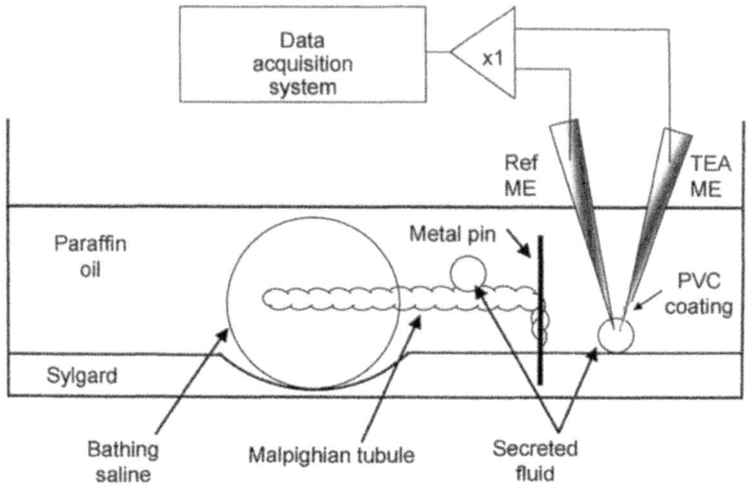

FIG. 11 Use of TEA-selective microelectrodes for measurement of the concentration of salicylate in droplets of fluid secreted by isolated Malpighian tubules in the Ramsay assay. An isolated pair of Malpighian tubules is placed in a droplet of bathing saline under paraffin oil. One Malpighian tubule remains in the saline, and the other is pulled out and wrapped around a stainless steel pin embedded in the Sylgard-line base of a Petri dish. Secreted fluid droplets collect at the ureter which is positioned just outside the bathing saline droplet. Secreted fluid droplets are collected on glass rods and placed on the bottom of the dish adjacent to calibration droplets containing known concentrations of TEA in *Drosophila* saline. For each droplet, the potential difference between the TEA-selective microelectrode (TEA ME) and the reference microelectrode is measured by a unity gain high impedance ($>10^{15}$ Ω) operational amplifier. Voltages are digitized and recorded on a PC-based data acquisition system. TEA concentration in the secreted droplets is calculated from the voltage difference between the secreted droplet and the calibration droplets. Redrawn from O'Donnell and Rheault (2005, Fig. 2), with modifications. Reprinted with permission from the Company of Biologists.

segments bathed in 5 μM TEA, TEA must be actively transported into the secreted fluid against an opposing electrochemical gradient.

Active secretion of TEA by the lower tubule (Figs. 13, 14) is accompanied by acidification of the urine, active secretion of Ca^{2+} and reabsorption of K^+ and Cl^- by this segment (O'Donnell and Maddrell, 1995). A possible mechanism for TEA secretion across the apical membrane of the lower tubule is exchange of organic cations for protons (Bijelic and O'Donnell, 2005), as proposed for organic cation transport across the apical membrane of vertebrate renal tissues (Pritchard and Miller, 1991).

The mechanism of TEA transport across the basolateral membrane of *Drosophila* tubules was studied using measurements of basolateral

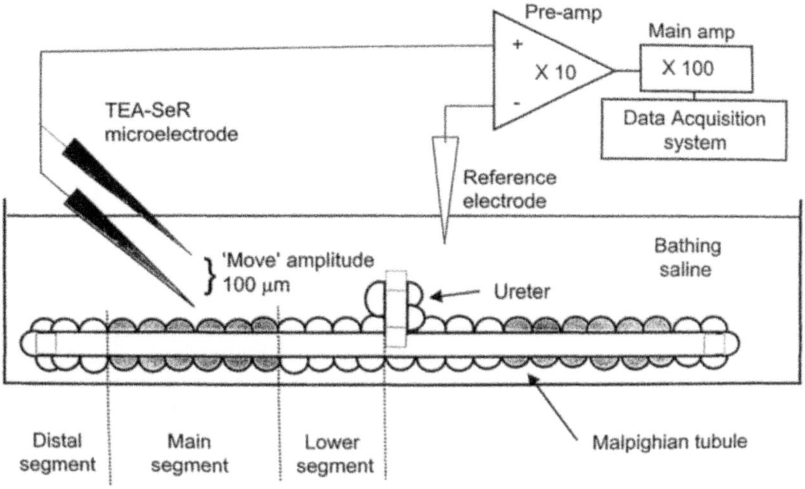

FIG. 12 The arrangements for recording tetraethylammonium (TEA) flux using TEA-selective microelectrodes based on the ion exchanger Corning 477317. The tip of a TEA-selective self-referencing microelectrode (TEA-SeR) microelectrode is moved from a position 10 μm away from the surface of the Malpighian tubule epithelium to a position 100 μm farther away by computer-controlled stepper motors. The differential signal between the two positions is amplified 10 × in the head stage and a further 100 × in the connected amplifier for a total voltage amplification of 1000 × . An associated PC-based data acquisition system running ASET software records the voltages and controls the stepper motors for microelectrode positioning and movement. The relative locations of the three segments of an anterior pair of Malpighian tubules and the connecting common ureter are indicated. Redrawn from O'Donnell and Rheault (2005, Fig. 1), with modifications. Reprinted with permission from the Company of Biologists.

membrane potential (V_{bl}) and uptake of [^{14}C]-labelled TEA (Rheault *et al.*, 2005). In common with vertebrate renal tissues, uptake of [^{14}C]-labelled TEA is dependent on metabolism and saturable. TEA uptake in *Drosophila* tubules is inhibited by a number of other type I organic cations (e.g. other quaternary ammonium compounds, cimetidine, quinine) and is dependent on basolateral membrane potential. Conditions that depolarize basolateral membrane potential (e.g. high K^+ saline) reduce TEA uptake, whereas hyperpolarizing conditions (low K^+ saline) increase uptake. Addition of TEA to the saline bathing Malpighian tubules rapidly depolarizes the basolateral membrane potential in lower Malpighian tubules of *Drosophila*, indicating that TEA uptake is electrogenic. However, the effects of TEA on basolateral membrane potential are not altered when K^+ channels are first blocked with Ba^{2+}, indicating that TEA uptake does not occur through K^+ channels in the lower tubule (Fig. 15).

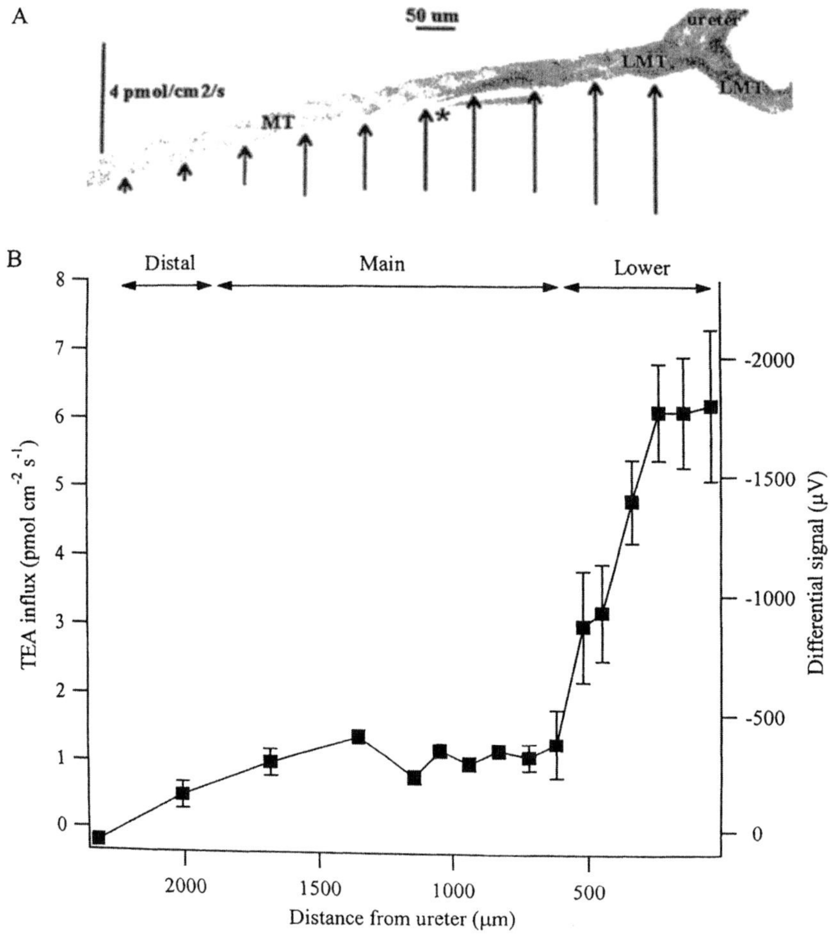

FIG. 13 (A) Representative scan of tetraethylammonium (TEA) flux at locations along the secretory segment of the Malpighian tubule (MT) and the lower MT (LMT) of *Drosophila melanogaster*. Tubules were bathed in saline containing 100 μM TEA. The common ureter and part of the other LMT of the pair are shown. The tip of the TEA-SeR microelectrode is located just above the asterisk. The image is a collage formed from two images. At each site, indicated by arrowheads, ASET software calculated the TEA-specific signal differences (ΔV; μV) between the two limits of microelectrode excursion by subtracting the voltage at the outer limit of the excursion from that measured at the inner limit. The length of each arrow corresponds to the magnitude of TEA influx. (B) TEA influx as a function of distance from the ureter along lower, main and distal segments of the MT. An influx of TEA reduces TEA concentration in the unstirred layer adjacent to the surface of the tissue, and the corresponding voltage difference is therefore negative. Distance 0 on the abscissa corresponds to the junction of the ureter and the LMT. Both the differential signal recorded by the TEA-SeR microelectrode (right ordinate) and the calculated TEA influx (left ordinate) are shown ($N = 4$). Redrawn from Rheault and O'Donnell (2004, Fig. 1). Reprinted with permission from the Company of Biologists.

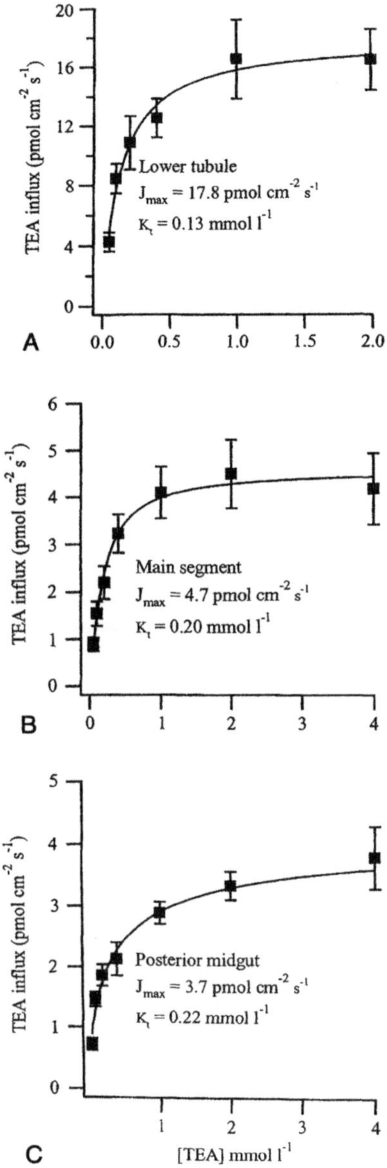

FIG. 14 Concentration–response curves for tetraethylammonium (TEA) influx in the (A) lower Malpighian tubule (LMT) of *Drosophila melanogaster*, (B) Malpighian tubule main segment and (C) midgut. Each point is the mean ± S.E.M. of $N = 4$–7 preparations. Values for J_{max} and K_t were determined by non-linear regression analysis. Redrawn from Rheault and O'Donnell (2004, Fig. 4). Reprinted with permission from the Company of Biologists.

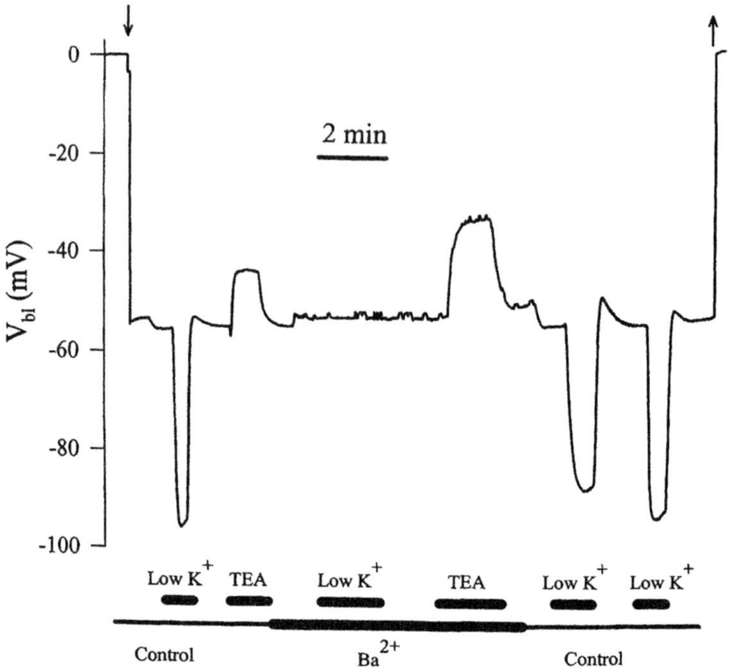

FIG. 15 Recording showing the effects of 6 mM Ba^{2+} on basolateral membrane potential (V_{bl}) of the lower Malpighian tubule of *Drosophila melanogaster*. The tubule was bathed in control saline ($[K^+] = 20$ mM), low K^+ saline (2 mM) and control saline containing 1 mM TEA. The duration of exposure to each saline is indicated by the horizontal bars (Rheault *et al.*, 2005, Fig. 7). Reprinted with permission from the American Physiological Society.

Addition of verapamil, a type II organic cation, also causes a concentration-dependent inhibition of TEA uptake. Previous studies on snake renal proximal tubules have demonstrated that verapamil competes with TEA for the basolateral organic cation uptake pathway (Kim and Dantzler, 1995). Moreover, the plant alkaloid, nicotine, a demonstrated P-gp substrate in insect Malpighian tubules, inhibits basolateral TEA uptake. Taken together, these results suggest a broad overlap in specificity of the basolateral organic cation transport pathway for type inorganic cations such as TEA and type II organic cations such as nicotine and other alkaloids. Studies in mammals also indicate that type II OC's inhibit transport of type I OC's, but not *vice versa* (Wright and Dantzler, 2004).

Secretion of TEA by Malpighian tubules appears to be common but not universal among insects. TEA is also transported from bath to lumen in Malpighian tubules of seven species of insect from six orders. Based on the ratio of secreted [TEA] to that in the bathing medium, as well as the value of the transepithelial potential recorded in earlier studies by several groups,

Rheault *et al.* (2006) concluded that TEA is actively secreted by tubules of the orthopterans *L. migratoria*, and *Ac. domesticus*, the dictyopteran *P. americana*, the coleopteran *Te. molitor*, the lepidopteran *Trichoplusia ni* and the hemipteran *Oncopeltus fasciatus*. By contrast, Malpighian tubules of two blood-feeders, the hemipteran *R. prolixus* and the dipteran *Ae. aegypti* do not actively transport TEA under normal physiological conditions. Transport of TEA but not nicotine increases during the moult in the Malpighian tubules of *Rhodnius*, but the concentrations of TEA in the secreted fluid are still consistent with passive accumulation in response to the lumen-negative transepithelial potential. Both of the latter species actively transported the type II organic cation nicotine. The absence of active transport of the prototypical type I organic cation TEA by the Malpighian tubules of these species is puzzling and suggests that toxic levels of type I organic cations in the hemolymph may be avoided by excretion across other epithelia such as the midgut, as observed in *Drosophila* (Rheault and O'Donnell, 2004). Alternatively, type I OCs in *Rhodnius* and *Aedes* may be metabolized into other compounds, which are either non-toxic or are readily excreted by the tubules or other epithelia.

Several types of evidence suggest that there are separate pathways for transport of the type I organic cation TEA and the type II organic cation nicotine in insect Malpighian tubules. Verapamil reduces TEA transport by the lower tubule but not main segment of *Drosophila* Malpighian tubules (Rheault and O'Donnell, 2004). Inhibition by verapamil suggests the additional involvement of a P-gp-like transport mechanism for TEA in the lower tubule that is not involved in the secretion of TEA by the main segment.

The possibility that excretory mechanisms are altered by exposure to dietary toxins was examined by feeding *Drosophila* larvae a diet enriched in TEA (Bijelic *et al.*, 2005). More than 75% of third instar *Drosophila* larvae survive on a diet containing 100 mM TEA. There is a synergistic, rather than additive, increase in larval mortality when the organic cations quinidine or cimetidine are added to a diet containing 100 mM TEA; the combined response was significantly greater than the sum of the mortalities associated with TEA and either cimetidine or quinidine alone. The synergism may reflect the competitive inhibition of TEA transport by either quinidine or cimetidine, resulting in higher hemolymph levels of TEA. The rate of decline in hemolymph TEA after transfer to TEA-free diet is reduced by 58% when larvae are maintained on diet that contains 10 mM cimetidine and 100 mM TEA as compared to the rate of decline when the diet contains 100 mM TEA alone. Cimetidine presumably slows the rate of decline of the TEA concentration in the hemolymph because it reduces the rate of TEA secretion by transporters in the Malpighian tubules and posterior midgut. When larvae are fed a diet that contains $\leq$20 mM TEA and are subsequently transferred to TEA-free diet, the concentration of TEA in the hemolymph declines at a rate that can be accounted for by active and saturable transport of TEA across the Malpighian tubules and the midgut. However, when larvae are transferred from a diet containing 100 mm TEA to a TEA-free diet, hemolymph TEA concentration declines at least 10 times more rapidly than predicted on the basis of active, saturable transport across these tissues. Passive transport into the gut may provide a non-saturable means for rapid clearance of hemolymph TEA when animals are fed TEA-free diets (Bijelic *et al.*, 2005). The gut contents may thus act as a passive 'sink' for toxins in the hemolymph, providing that the insect can feed upon toxin-free foods after ingestion and absorption of high levels of toxins.

Transepithelial TEA secretion and secreted fluid TEA concentration increase significantly in tubules isolated from third instar larvae that are maintained for 24 h on media containing 50 or 100 mM TEA (Bijelic *et al.*, 2005). This increase suggests an adaptive response, whereby ingestion of a TEA is associated with up regulation of TEA excretion by the Malpighian tubules. The effects of long-term exposure ($\sim$ 10 d) to another toxin, the organic anion salicylate, are discussed in Section 8.3.

8.2 TRANSPORT OF TYPE II ORGANIC CATIONS BY P-GLYCOPROTEIN-LIKE MECHANISMS

P-gps function in MDR in humans by acting as ATP-dependent drug efflux pumps to maintain the intracellular concentrations of anticancer drugs below the cytotoxic levels. Expression of P-gp on the secretory surfaces in a number of tissues including adrenal gland, kidney, liver, intestinal tract and uterine epithelium also suggests a role either in transporting substances away from tissues and out of the body or in decreasing absorption from the surroundings. P-gps in aquatic species aid survival in environments which contain high levels of anthropogenic pollutants or toxins produced by other organisms. The term multixenobiotic resistance (MXR) is used to distinguish this phenomenon from MDR in humans (Bard, 2000).

Substrates of P-gps are usually moderately hydrophobic planar molecules, including type II organic cations (Wright and Dantzler, 2004). In addition, P-gps transport relatively hydrophobic compounds that are uncharged or neutral, including digoxin, colchicine and some corticosteroids. It is generally thought that P-gps function as 'flippases' to flip substrates from the inner to the outer leaflet of the plasma membrane against an intramembrane concentration gradient. The substrate then diffuses out of the outer leaflet into the extracellular medium (Higgins and Gottesman, 1992). In the sections below, the possible roles of P-gps in transporting insecticides are discussed first, followed by a section dealing with evidence for P-gps in specific insect tissues such as the Malpighian tubule. The subsequent sections deal with the chemical ecology of P-gp substrates such as nicotine and other alkaloids.

8.2.1 *P-glycoproteins and insecticide effects*

The P-gp inhibitor verapamil (100 μM) increases by 1.5 to 2-fold the toxicity of examples of three classes of insecticides (cypermethrin, endosulfan and ivermectin) in three strains of *Culex* mosquitoes but has no effect on the toxicity of the organophosphate chlorpyrifos (Buss *et al.*, 2002). Although verapamil is best known as a potent blocker of L-type calcium channels, it is also a well known example of agents known as chemosensitizers, which have been shown to reverse multidrug resistance. Although relatively non-toxic themselves, chemosensitizers interact with P-gp so as to increase the toxicity of compounds such as insecticides. Two of the insecticides also strongly inhibit the efflux of a fluorescent P-gp substrate out of CH1 cells, a standard test of P-gp inhibition. The organophosphate insecticides fenthion and chlorpyrifos and the organochlorine DDT have no effect on R123 efflux, whereas ivermectin and the cyclodiene organochlorine endosulfan reduce efflux. Ivermectin is the most potent P-gp inhibitor in these assays. Widely used as an antihelminthic and as a commercial insecticide, ivermectin is a semi-synthetic analogue of avermectin, which is produced by the actinomycete *Streptomyces avermitilus*. Ivermectin is well known as a P-gp inhibitor in studies of mammalian cells and tissues as well as clinically.

The interactions of insecticides with P-gps have also been studied in cultured murine melanoma cells transfected with a human MDR gene (Bain and LeBlanc, 1996). Six out of eight organophosphate insecticides, including chlorpyrifos, inhibit P-gp, as measured by the inhibition of efflux of the P-gp substrate doxorubicin. Although carbamate and pyrethroid insecticides do not interact with P-gp, many organophosphorus and organochlorine pesticides do bind to the transporter. However, many pesticides that are capable of binding to P-gp are not transported by it. Based on structure-activity analyses, pesticides that are good candidates for P-gp inhibitors have lipophilicity (measured as the log of the octanol: water partition coefficient, K_{ow}) of between 2.5 and 4.5 and a molecular mass of 300 or more (Bain and LeBlanc, 1996). Using these criteria, pesticides likely to inhibit P-gp are ranked in the order endosulfan > chlorpyrifos > DDT > fenthion. Although chlorpyrifos is not an inhibitor of P-gp based on reduction in R123 efflux from CH1 cells (Buss *et al.*, 2002), chlorpyrifos increases P-gp expression in rats and the active metabolite chlorpyrifos oxon inhibits P-gp substrate binding. Differences in the effectiveness of chlorpyrifos as a P-gp inhibitor may relate to differences in substrate specificities in P-gps derived from expression of different MDR genes (Buss *et al.*, 2002).

The significance of P-gp interaction with commercial pesticides is twofold. Firstly, P-gp may be a component of insecticide resistance, especially given its potential to confer resistance against a wide range of pesticides. This adds a complication to control programes for mosquitoes and possibly other disease vectors and agricultural pests. However, the finding that insecticide toxicity is increased by chemosensitizers such as verapamil opens up a new avenue for development of novel, environmentally benign insecticide synergists. In the study of Buss *et al.* (2002) for example, verapamil was itself non-toxic but increased the toxicity to examples of three insecticide classes. The idea is that compounds of low toxicity themselves may permit reduced insecticide application rates because they block elimination of insecticides from pest species.

8.2.2 *P-glycoproteins in the gut, Malpighian tubules and anal papillae*

Both the gut and the Malpighian tubules are important sites of P-gp expression in insects. In *D. melanogaster*, P-gps are the products of expression of three genes, MDR49, MDR50 and MDR65. Larvae of *D. melanogaster* fed on diets containing a non-toxic level of the P-gp substrate colchicine (10 μM) show increased expression of MDR genes in brain and gut relative to control larvae raised on colchicine-free diet (Tapadia and Lakhotia, 2005). Increased MDR expression is evident in all sections of the gut (foregut, midgut, hindgut and Malpighian tubules).

Prompted by evidence that promoters of human multidrug resistance MDR genes harbour stress-responsive elements, which respond to heat shock, heavy metals and cytostatic drugs, Tapadia and Lakhotia (2005) also studied expression of *Drosophila* MDR genes under heat stress. After heat shock at 37 °C for 40 min, third instar larvae were either immediately dissected and different tissues were fixed or were allowed to recover from heat shock for different time periods at room temperature before tissue fixation. A 40-min heat shock significantly increases the expression of MDR genes in different larval tissues such as the brain and salivary glands. Levels of the MDR49 transcripts remain elevated above control levels for more than 60 min after heat shock, and expression returns to basal non-heat shock levels only after 90 min recovery. By contrast, the levels of MDR65 transcripts are unaffected by heat shock.

These differences in the effects of heat shock on expression of MDR49 versus MDR65 are explicable on the basis of differences in the actions of the heat shock transcription factor (HSF), which has been widely implicated in the induction of the heat shock genes in eukaryotes by heat and other forms of stress. The involvement of HSF in the up regulation of MDR genes in *Drosophila* larvae has been examined after heat shock by immunostaining polytene chromosomes with anti-HSF antibody. Although HSF is found at the MDR49 gene locus on polytene chromosomes from heat shocked salivary glands, it is not present at the 65A region, where the MDR65 gene is located, either in control or in heat-shocked salivary glands. These findings are consistent with a role for HSF in mediating the increased expression of MDR49 relative to MDR65 in response to heat shock. MDR genes use alternative promoters in response to different stimuli in human cell lines and tissues and there are distinct nuclear protein-binding sites in the promoter of the murine *MDR* gene. It will be of interest to analyze promoter regions of the *MDR* genes of *Drosophila* to further examine differences in the regulation of these genes.

Several lines of evidence indicate the presence of a P-gp-like pump in the anal papillae of the larvae of the non-biting midge *Ch. riparius* (Podsiadlowski *et al.*, 1998). The anal papillae are stained *in vivo* by the fluorescent P-gp substrate rhodamine 123 in the presence of the P-gp inhibitor verapamil. There is no staining in the absence of the inhibitor, consistent with the role of a P-gp-like pump in transporting rhodamine 123 out of the cells. Immunohistochemical analysis also reveals P-gp-like molecules in the anal papillae. Positive staining is evident in the anal papillae when a monoclonal antibody against human P-gp is used in conjunction with secondary antibodies conjugated to peroxidase. There is no staining when non-immune mouse serum replaces the primary antibody, or when antibodies against human MRP are used (Podsiadlowski *et al.*, 1998). Hydrolysis of ATP in homogenate fractions from the posterior segments (including the anal papillae) of *Chironomus* larvae is enhanced when P-gp substrates such as vinblastine and ivermectin are present. By contrast, addition of the MRP substrates arsenate and glutathione does not enhance ATP consumption. Hydrolysis of ATP by other ATPases was blocked in these experiments by addition of a cocktail containing appropriate inhibitors such as ouabain for the Na^+/K^+-ATPase. Lastly, synergism between the toxicity of the P-gp substrate ivermectin and P-gp inhibitors is consistent with a role for P-gp in protection of the larvae against P-gp substrates such as ivermectin. The LD_{50} of ivermectin was reduced nearly threefold by sublethal concentrations of either of the P-gp inhibitors verapamil or cyclosporin.

A P-gp-like mechanism in the Malpighian tubules of the tobacco hornworm is responsible for active excretion of nicotine and other alkaloids (Gaertner *et al.*, 1998). Malpighian tubules were cannulated and P-gp substrates were applied to the bathing fluids or to both the bathing fluids and the lumen. The lumenal contents were flushed out at the end of the experiment and lumen to bath ratios of the substrates were measured. Lumen to bath ratios of 4.0, 10.1 and 3.2 were measured at bathing saline nicotine concentrations of 0.05, 0.5 and 5 mM, respectively. Application of the P-gp inhibitor verapamil (0.025 mM) reduced the lumen to bath ratio from 10 to 1.2 for tubules in saline containing 0.5 mM nicotine. Stirring the bathing saline to minimize unstirred layer gradients (i.e. depletion of nicotine in the unstirred layer at the basolateral surface) increased the lumen to bath ratio from 4 to 12.3. Body movements tend to minimize unstirred layer gradients *in vivo* (cf. Collier and O'Donnell, 1997), so the highest lumen to bath ratios are probably indicative of the maximal capacity of the tubules to secrete nicotine. Malpighian tubules of the tobacco hornworm are immunopositive for P-gp, as is the blood–brain barrier (Murray *et al.*, 1994). P-gp in the latter tissue thus protects the nervous system from the effects of nicotine until the toxin can be secreted into the lumen of the Malpighian tubule and eliminated.

It is difficult to estimate the contribution of P-gp to elimination of toxins by the Malpighian tubules using only measurements of lumen to bath ratios. The problem is that changes in the rates of fluid secretion rate influence the concentration of the solute in the lumen, particularly if the permeability of the tubule wall to the solute is high. A stimulant of fluid secretion will thus appear to inhibit transport of a P-gp substrate if only lumen to bath concentration ratios are measured, since the concentration of the solute in the lumen will be reduced. However, the transepithelial flux of the solute may actually increase, since the reduction in lumenal concentration will reduce the extent passive lumen to bath diffusion of the solute. It is thus essential to measure both the concentration of the transported solute in the lumen and the fluid secretion rate, so that the transepithelial flux may be calculated. For tubules of most species, fluid secretion rates are usually expressed in units of nanoliters per minute and concentrations of solutes in the lumen as μM. Transepithelial flux is calculated as the product of fluid secretion rate (nl min^{-1}) and luminal solute concentration (μM) and is usually expressed in units of fmol min^{-1} for *Drosophila* tubules.

Measurement of fluid secretion rate permits non-specific toxicity of P-gp inhibitors or of high concentrations of P-gp substrates themselves to be distinguished from specific inhibition of transepithelial flux through direct effects on P-gps or other types of transporter. A study of transepithelial flux of several fluorochromes revealed inhibition of transepithelial flux of the P-gp substrates rhodamine 123 and daunorubicin by verapamil (Leader and O'Donnell, 2005). The results of the latter study also indicate that rhodamine 123 is not an appropriate substrate for analysis of P-gp function in the tubules of species such as *Drosophila* because it dramatically inhibits fluid secretion. Measurements of only lumen-to-bath ratios of rhodamine 123 would have led to profound errors in estimates of the kinetic parameters (J_{max}, K_t) for rhodamine 123 transport. Rhodamine 123 is also used as a mitochondrial dye, so its effects on cell metabolism and hence on fluid secretion may be particularly apparent in the highly active cells of Malpighian tubules. Rhodamine 123 inhibits fluid secretion by tubules of crickets (*Teleogryllus commodus*) and *Drosophila* at concentrations of 10 and 0.1 μM, respectively. The latter value is nearly

40-fold less than the concentration at which rhodamine 123 transport is half maximal (K_t) in cricket tubules. Analysis of dye transport solely by image analysis of the cellular and lumenal compartments would not reveal this additional information about the effects of the dye on cell function. The use of rhodamine 123 is also complicated by the finding that principal cells of dipteran tubules retain rhodamine 123 for a short period and the stellate cells reabsorb the dye from the lumen. The dye accumulates in stellate cell mitochondria and intensely fluorescent vesicles which are probably lysosomes (Meulemans and De Loof, 1992).

A novel fluorometric method for the analysis of transepithelial fluxes of fluorescent dyes, including some P-gp substrates, has been developed by Leader and O'Donnell (2005). The method is also suitable for use with fluorescent substrates of other transporters for organic cations or organic anions. Isolated tubules are setup in Ramsay secretion assays and nanoliter samples of secreted fluid are collected in optically flat rectangular glass capillaries. The measurement of fluorochrome concentration in the secreted droplets circumvents many of the problems associated with confocal laser scanning microscopy (CLSM) of the isolated tubules. The lumen and the cells are often difficult to distinguish by CLSM because of the presence of numerous opaque concretions and extensive apical microvilli. Problems with confocal laser scanning microscopy of fluorochrome transport of isolated tissues were first noted by Gaertner and Morris (1999). The fluorescent P-gp substrates rhodamine 123 and daunorubicin did not appear to accumulate in the lumen in confocal laser scanning micrographs, in contrast to the results of an earlier study demonstrating lumen to bath ratios >1 for the radiolabelled P-gp substrates nicotine and vinblastine (Gaertner *et al.*, 1998). Gaertner and Morris (1999) suggested that the apparent lack of transport of the fluorescent P-gp substrates may have been the result of quenching, reflection or absorption of the exciting and/or emitted light by the tissues and by opaque uric acid crystals present in the lumen. Similarly, it was not possible to clearly visualize fluorochromes and the lumen of many of the cricket Malpighian tubules and all *Drosophila* tubules examined because of opaque concretions in the cells and/or lumen (Leader and O'Donnell, 2005). Additional concerns are that concretions move in the lumen during fluid secretion, thereby altering the amount of laser light passing through the lumen, and that concretions may be transferred from the cell to the lumen, particularly when fluid secretion is stimulated (Hazelton *et al.*, 2001). By contrast, collection of secreted fluid samples in optically flat glass capillaries and subsequent analysis by CLSM avoids the errors associated with the presence of concretions, which sink to the bottom of the secreted fluid droplet and are not collected in the capillary. Estimates of dye concentration are less affected by variations in cytoplasmic pH and the calibration solutions can be adjusted to resemble closely the pH and ionic strength of the secreted fluid. Measurement of dye transport by the method of Leader and O'Donnell (2005) also provides a faster, cheaper and more sensitive method than the use of radiolabelled substrates.

The evidence for P-gps in insect Malpighian tubules and the finding that nicotine is transported by a P-gp-like mechanism suggests that P-gps may be involved in excretion of other alkaloids as well. Alkaloids are nitrogenous bases which protect the plants that produce them from attack by insects and other herbivores. In turn, insects have evolved means to reduce alkaloid toxicity. In many species of insect which feed upon alkaloid-rich plants, the Malpighian tubules remove alkaloids such as nicotine, morphine and atropine at high rates (Maddrell and Gardiner, 1976). Nicotine is transported unchanged in *Rhodnius* and *Pieris*, whereas it is transformed into an unidentified substance in *Musca* and *Calliphora*. Transport occurs at high rates, and large lumen-hemolymph concentration differences are established. Transport is saturable ($J_{max} = 700\,\text{pmol min}^{-1}$ tubule, $K_t = 2$–$3\,\text{mM}$ in *Rhodnius*) and is not coupled to ion transport. Competition studies suggest that a common transport mechanism is used for nicotine, morphine and atropine. Nicotine transport by *Rhodnius* tubules is unaffected by organic anions, such as amaranth and benzyl penicillin, or by the organic anion transport inhibitor, probenecid. In vertebrates, atropine and morphine are also transported by the same renal transport system which handles organic bases such as *N*-methylnicotinamide, choline and histamine. By contrast, nicotine and TEA appear to be substrates of different transporters in *Rhodnius* and other insects (Rheault *et al.*, 2005).

8.2.3 *Roles of metabolism and excretion in adaptation of Lepidoptera to nicotine-rich diets*

Nicotine tolerance is well known for *Ma. sexta* and for several other sphingids of the subfamilies Macroglossinae and Sphinginae, whereas members of the subfamily Smerinthinae are sensitive to poisoning by nicotine (Wink and Theile, 2002). A simple target-site modification as a basis for nicotine tolerance has been ruled out based on the absence of apparent amino acid substitution in the putative nicotine binding site of the f-subunits of nicotinic acetylcholine receptors from nicotine-tolerant and nicotine-sensitive sphingids. This conclusion is supported by the results of feeding experiments which show that larvae of *Ma. sexta* and other sphingids of the Macroglossinae and Sphinginae tolerate not only nicotine, but also many other alkaloids that affect neuroreceptors other than acetylcholine receptors.

When nicotine is injected into larvae of nicotine-tolerant taxa, 80–90% is apparently converted to polar conjugates or degradation products such as nicotine *N*-oxide or cotinine and only 10–20% is recovered as free nicotine

(Wink and Theile, 2002). Usually more than 98% of the recoverable alkaloids are found in the faeces and excretion peaks <6 h after injection in tolerant taxa. Rearing larvae of *Ma. sexta* on a nicotine-rich diet, results in higher nicotine degradation (Snyder *et al.*, 1994) and faster nicotine elimination relative to naïve larvae (Fig. 16). Application of the cytochrome P450 inhibitor SKF 525A (proadifen) reduces the formation of nicotine *N*-oxide and the rate of alkaloid degradation. Sphingidae are thus able to live on host plants rich in otherwise toxic secondary

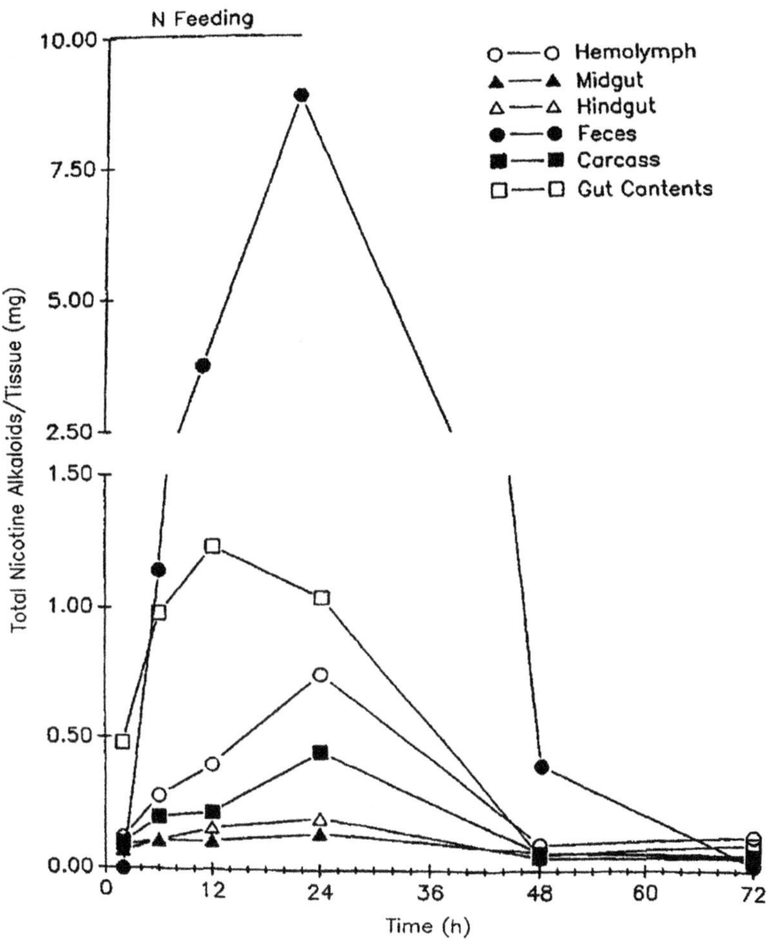

FIG. 16 Levels of nicotine and its metabolites in tissues of fifth instar larvae of *Manduca sexta.* The larvae were fed diet containing 0.75% nicotine diet for 24 h (0–24 h) and were then returned to a nicotine-free diet (24–72 h; Snyder *et al.*, 1994, Fig. 1). Reprinted with permission of Elsevier Science.

metabolites by the combined actions of an inducible detoxification mechanism and rapid and inducible excretion.

The high capacity for excretion of alkaloids is clearly evidenced by the results of Self *et al.* (1964), Maddrell and Gardiner (1976) and Snyder *et al.* (1994). Competition between morphine, atropine and nicotine indicates a broad specificity of the alkaloid transporter in Malpighian tubules (Maddrell and Gardiner, 1976). It seems likely that the same mechanism would also actively transport the nicotine metabolites nicotine-*N*-oxide and cotinine-*N*-oxide, thus resulting in a rapid clearance from the insect (Snyder *et al.*, 1994).

It has been suggested by Snyder *et al.* (1994) that active transport of alkaloids or more generally high excretion rates are not sufficient, and perhaps not necessary, to explain the adaptation to nicotine in larvae. Malpighian tubules of *Pieris brassicae* larvae also actively transport nicotine at very high rates (Maddrell and Gardiner, 1976), yet this insect is not tolerant of dietary nicotine (David and Gardiner, 1953). However, the study of Murray *et al.* (1994) suggests a possible explanation for differences of nicotine toxicity to *Manduca* and *Pieris*. P-gp-like transporters in the blood brain barrier of *Manduca* protect the central nervous system from the effects of nicotine (Murray *et al.*, 1994). These authors suggest that detoxifying enzymes as well as the nicotine pump are likely to account for the metabolic blood–brain barrier to nicotine. In this scenario, then, the sensitivity of *Pieris* to nicotine may reflect the absence of an efficient blood–brain barrier toward nicotine and its metabolites. The fate of nicotine in *Ma. sexta* appears to resemble a general mechanism for handling allelochemicals in the Lepidoptera. In other insects that are unaffected by chemicals such as xanthotoxin, precocene II, or α-terthienyl, tolerance also involves rapid metabolism followed by excretion (Haunerland and Bowers, 1985; Iyengar *et al.*, 1990; Nitao, 1990).

8.2.4 *What other compounds are transported by P-glycoprotein-like mechanisms in insects?*

Insects which feed on conifer needles are also exposed to high levels of piperidine alkaloids that may be suitable substrates of P-gps. Spruce and pine needles contain ~0.2% wet weight of several disubstituted piperidine alkaloids, some of which have insecticidal or teratogenic activity. Although some beetles sequester such compounds as a defence against predation by other insects and vertebrates, they are accumulated in the tissues of western spruce budworm larvae (*Choristoneura occidentalis*) feeding on the needles (Kamm *et al.*, 1998). The frass of the larvae contain the same alkaloids found in the needles and also some more polar metabolites, indicating processing of these compounds and excretion of metabolized and unaltered forms. The absence of sequestration of deterrent or toxic alkaloids is consistent with the cryptic coloration of spruce budworms, in contrast to aposematic coloring of beetles that synthesize some of the same alkaloids *de novo* (Kamm *et al.*, 1998).

P-gps have also been implicated in resistance to the toxin α-amanitin, a compound found in mushrooms which are fed on by members of the *Drosophila quinaria* species group (Begun and Whitley, 2000). Toxicity of α-amanitin is due to interference with RNA polymerase activity. Although *D. melanogaster* feed on fruit and are unlikely to be exposed to α-amanitin, some strains are resistant to the toxin and genetic studies indicate that polymorphism in a MDR gene (MDR65A) is implicated in resistance. However, DNA sequence analysis does not show any differences in the amino acids coded for by the MDR65A genes of α-amanitin sensitive and resistant strains (Begun and Whitley, 2000). Given that MDR in mammalian cells is often the result of amplification of MDR genes, the latter authors suggest that a mutation in gene regulation may be the basis for α-amanitin resistance in *D. melanogaster*. The mutation might result in either increased expression of a P-gp or its modification by other molecules such as protein kinases.

P-gps may also be involved in excretion of some types of terpenoids in insects, although direct evidence is lacking. The neem tree is grown in most tropical and sub-tropical areas of the world for shade, reforestation and as a source of natural insecticides and medicines. Toxic effects in insects are produced primarily by azadirachtin, a complex tetranortriterpenoid limonoid from the neem seeds. Terpenoids are a large and structurally diverse class of naturally occurring organic compounds formed from 5-carbon isoprene units which are often arranged to form multicyclic structures. In addition to azadirachtin, terpenoids include menthol, camphor and the cannabinoids and they contribute to the flavours of cinnamon and cloves and the scent of eucalyptus. Terpenoids inhibit transport of digoxin by P-gps in a dose-dependent manner and the IC_{50} values in cultured mammalian cells are near or below the value for the commonly used inhibitor verapamil (Yoshida *et al.*, 2006).

Azadirachtin acts as an antifeedant and repellent in insects and it also interferes with normal development by inhibiting the release of prothoracicotropic hormones and allatotropins (Mordue and Blackwell, 1993; Williams and Mansingh, 1996). Tritiated dihydroazadirachtin A, fed in a blood meal to *R. prolixus*, is absorbed, transported into the hemolymph and then excreted unmetabolized by the organism (Garcia *et al.*, 1989). The level of dihydroazadirachtin A in the hemolymph rises steeply to a peak 24 h after feeding, then declines gradually as it is excreted. The highest amount per milligram tissue is found in the Malpighian tubules and midgut, the lowest in the head and the remaining body. Unchanged labelled compound appears in the urine, consistent with excretion of unmetabolized dihydroazadirachtin A through the Malpighian tubules.

It will be of interest in the future to determine if other terpenoids are transported by insect Malpighian tubules and which transporters are involved.

8.2.5 *Transport of cardiac glycosides and miserotoxin*

The milkweed bug *O. fasciatus* feeds exclusively on milkweeds, which contain cardiac glycosides as a defence against vertebrate and invertebrate herbivores. Toxicity of cardiac glycosides results from inhibition of the Na^+/K^+-ATPase. Cardiac glycosides are composed of a sugar (glycoside) conjugated to an aglycone (steroid) moiety and are divided into two classes: cardenolides, which have an unsaturated butyrolactone ring, and the bufadienolides, which have an α-pyrone ring. The toad venom bufalin is a well-known bufadienolide and examples of cardenolides are ouabain, digoxin and digitoxin. Milkweed bugs tolerate diets rich in cardiac glycosides by removing these compounds from the hemolymph and sequestering them in the dorsolateral space (Duffey and Scudder, 1974). About 60–95% of the cardenolide is stored in the dorsolateral space, with the remainder found in ventral metathoracic gland secretions of adults or middorsal abdominal gland fluid of larvae. Concentrations of cardenolides in the urine–faeces mixture or hemolymph are low. Sequestration not only permits the bugs to exploit milkweeds as a food source, but provides the warningly colored insect with a chemical defence against predators. For defensive purposes, droplets of dorsolateral space fluid are released from an orifice in each segment.

Physical sequestration of glycosides in special spaces presumably avoids toxicosis that would result from elevated hemolymph levels of these compounds. Sequestration is energy independent (i.e. insensitive to the mitochondrial uncoupler dinitrophenol) and follows first order kinetics (Duffey *et al.*, 1978). The sensitivity of sequestration is quite dramatic, and begins at hemolymph concentrations as low as 3.5×10^{-7} M of ouabain, for example. Moreover, sequestration of one cardenolide is independent of other cardenolides. Phase contrast microscopy reveals the presence of a distinct emulsion phase, and physical sequestration in emulsion particles is consistent with the observed energy independence, first order kinetics and absence of inhibition by presence of related compounds.

Polar cardenolides, such as ouabain, but not non-polar ones such as digitoxin, are sequestered. It appears that both preferential uptake and metabolism may be important in this regard. Digitoxin is converted to at least two more polar cardenolides, for example. Both polar and non-polar cardenolides cross the gut, and a protein carrier may transport them via the hemolymph to the special organs or glands of the dorsolateral space or other body regions.

The sequestration process is enhanced by a reduction in loss of cardiac glycosides through the excretory system (Meredith *et al.*, 1984). Malpighian tubules of *O. fasciatus* consist of two morphologically and physiologically distinct segments, both of which metabolize ouabain. In segment II, the distal segment, ouabain transport into the tubule lumen appears to be a passive phenomenon, since secretion does not occur against a concentration gradient. By contrast, segment I, the proximal segment, reabsorbs fluid and ouabain but not its metabolites. Ouabain reabsorption is independent of fluid secretion, and can occur against concentration gradients as large as 23-fold. Sodium concentration in the reabsorbed fluid (53 mM) is also elevated relative to the bathing saline (28 mM) or whole tubule secretion (9 mM), suggesting that ouabain reabsorption may be linked to sodium cotransport. More than 50% of ouabain is reabsorbed in slowly secreting tubules and as much as 93% is reabsorbed in rapidly secreting tubules. Reabsorption of ouabain by the tubules thus contributes to accumulation of glycosides in the dorsolateral space.

The grasshopper *Zonocerus variegatus* also feeds on plants containing cardiac glycosides (Rafaeli-Bernstein and Mordue, 1978), and in this case excretion rather than sequestration of the ingested toxins is important for survival. The tubules are passively permeable to ouabain, as are those of *Locusta*. However, exposure of *Zonocerus* to dietary ouabain for more than 12 days induces more effective excretion of the glycosides, resulting in lumen/hemolymph concentration gradients of radiolabelled ouabain (or its metabolites) exceeding 3. Although *Locusta* tubules are not poisoned by ouabain, other tissues account for the toxicity of ouabain given to whole animals.

The glycoside miserotoxin (3-nitro-1-propyl-β-D-glucopyranoside) is found in the diet (4.5–5.2% by dry weight) of the melanopline grasshoppers *Melanoplus bivittatus* and *Melanoplus sanguinipes* (Orthoptera: Acrididae). They are able to tolerate high levels of the glycoside, in part by excretion of the intact molecule. Miserotoxin is a causative agent in cattle poisoning on rangelands in the western United States (Johnson *et al.*, 2001), and was originally isolated from *Astragalus miser* var. *oblongifolius* (Leguminosae). It is also found in *As. miser* var. *serotinus* (timber milkvetch), which occurs in the southern interior of British Columbia. Other glycosides of 3-nitropropanol (NPOH) are also present in timber milkvetch. Miserotoxin is rapidly hydrolyzed by microbial enzymes of the rumen, and the aglycone is absorbed and converted to 3-nitropropionic acid (NPA) a potent inhibitor of mitochondrial enzymes essential to respiration. When the aglycone is administered to melanopline grasshoppers, detoxification is achieved by two routes: by oxidation of the aglycone to NPA which is then conjugated with glycine, and by glucosylation of the aglycone to miserotoxin, in each case followed by excretion (Johnson *et al.*, 2001). Karakin, a glucose triester of NPA is also found in *Astragalus*, and is detoxified after hydrolysis *in vivo*, conjugation of the NPA to the glycine, serine or glutamate, and excretion of the resulting amides (Majak *et al.*, 1998).

Very little is known of the molecular nature of cardiac glycoside transport by insect cells. In vertebrates, glycosides are transported by P-gps (e.g. Koren *et al.*, 1998). However, a member of the OATP family is responsible, at least in part, for transport of compounds such as ouabain by *Drosophila* Malpighian tubules (Torrie *et al.*, 2004). There are eight OATP genes in *Drosophila*, of which six are abundantly expressed in tubule, and of which three actively accumulate ouabain when expressed in S2 cells. *In vivo*, RNAi against just one of these genes (oatp58Db) reduces ouabain transport in *Drosophila* tubules by 50%. Colocalization of OATP and the Na^+/K^+-ATPase in the basolateral membrane of the principal cells of the tubule may explain the relative refractoriness of insect Malpighian tubules toward ouabain. In addition, transepithelial transport in Malpighian tubules is dependent upon the actions of the V-ATPase, as described in Section 2, with the Na^+/K^+-ATPase playing an ancillary or subordinate role.

8.3 TRANSPORT OF TYPE I ORGANIC ANIONS

Organic anions can be separated into two structural groups, based on molecular weight, net charge and hydrophobicity (Wright and Dantzler, 2004). Type I OAs are small (generally <400 Da) and typically monovalent compounds, such as *p*-aminohippurate (PAH), probenecid and fluorescein. Type II OAs are bulkier (generally >500 Da), frequently polyvalent and include, calcein, methotrexate and many glutathione, glucuronide and sulfate conjugates. Type I and type II organic anions are often preferentially transported by different transporters in vertebrate renal tissues (Wright and Dantzler, 2004), and this appears to be the case in insect Malpighian tubules as well.

The Malpighian tubules have long been known to accumulate acidic dyes that have been injected into the hemolymph (Lison, 1937). Acidic dyes include carboxylates and sulfonates, and are derivatives of more toxic compounds resulting from metabolism or xenobiotic compounds in plants which form part of the insect's diet. Insect Malpighian tubules actively excrete both acylamides, such as PAH acid, and sulfonates, such as indigo carmine and amaranth (Maddrell *et al.*, 1974). Transport of acidic dyes such as amaranth and indigo carmine into secreted fluid can be measured spectrophotometrically. For compounds such as PAH and benzyl penicillin K, C^{14}-radiolabelled compounds are used and the concentrations of the compounds in samples of fluid secreted by Malpighian tubules are measured by liquid scintillation spectrometry. Chromatographic analysis established that neither of these compounds is metabolized by the

Malpighian tubules. Acidic dyes such as indigo carmine are secreted more rapidly at higher bathing saline pH, suggesting that it is the concentration of the anionic form of the dye which determines the rate at which it is transported (Maddrell *et al.*, 1974). Although competition studies using isolated tubules of *R. prolixus* suggest separate transport systems for acylamides (such as PAH acid) and sulfonates (Maddrell *et al.*, 1974), analysis of fluorescein transport by cockroach tubules suggest one common carrier for organic acids (Bresler *et al.*, 1990).

Transport of type I organic anions such as fluorescein and PAH by insect Malpighian tubules requires the presence of sodium in the bathing saline. Sodium-dependence is also a characteristic of organic acid transport by proximal tubules of the vertebrate kidney. Although concentration gradients exceeding 1000 are accomplished for organic acid transport by the vertebrate proximal tubules, more modest gradients of 30-fold are established by the relatively permeable Malpighian tubules of *Calliphora* and 300-fold by the less permeable tubules of *Rhodnius* (Maddrell *et al.*, 1974). Comparisons can also be made on the basis of the J_{max}/K_t ratio, where J_{max} is the maximum rate of transport and K_t the Michaelis–Menten constant, corresponding to the substrate concentration at which transport is half maximal. J_{max}/K_t is the effective rate constant when the transport rate is proportional to the substrate concentration. This proportionality occurs when substrate concentrations are less then K_t, as occurs *in vivo* (Bresler *et al.*, 1990). For transport of the organic anion fluorescein, the value of J_{max}/K_t is 10.6 in locusts, higher than that in rats (2.72), fish (0.9) and turtles (0.3).

The extent to which organic anions are concentrated in the fluid secreted by tubules depends on several factors such as the passive permeability of the tubule wall and the rate of fluid secretion. An important aspect of the paper by Maddrell *et al.* (1974) is the explanation for differences in the relationship between fluid secretion rate and dye transport by tubules isolated from different species. In the blood-feeding hemipteran *R. prolixus* and the stick insect *Ca. morosus* dye transport is independent of fluid secretion rate. By contrast, dye secretion by the tubules of the blowfly *Calliphora* is very greatly affected by changes in the rate of fluid secretion, being approximately linearly related.

A classic method for estimating the passive loss of dye from lumen to bath over a wide range of fluid secretion rates was developed by Maddrell *et al.* (1974). In order to vary fluid secretion rate, the tubule lumen is cannulated and perfused. The isolated Malpighian tubule is placed in a droplet of saline under liquid paraffin and a pair of fine forceps mounted on a micromanipulator is brought close to the bathing drop. The distal end of the tubule is then looped round the nearer point of the forceps. The forceps are arranged so that their points open and close in the horizontal plane and the tips of the forceps are held together with a small clump.

A fluid filled cannula is connected to a motor-driven syringe through thick-walled plastic tubing. The cannula is mounted on a second micromanipulator and positioned so that it is coaxial with the lumen of the tubule. The cannula is then pushed through the wall of the tubule and on down the lumen until the taper of the cannula fills the lumen. Using this technique, the rate of perfusion of the tubule lumen can easily be varied > 40-fold.

Net dye transport by cannulated Malpighian tubules of *Calliphora* is low at low rates of fluid perfusion, increases with the rate of lumenal perfusion below about $10\,\mathrm{nl\,min^{-1}}$ but is not much affected by further increases in the rate of perfusion (Maddrell *et al.*, 1974). When the rate of perfusion is low, measurements of samples of fluid emerging from the open end of the tubule reveal that the concentration of dye in the secreted fluid is many times higher than both that of the bathing fluid and that of the fluid perfused at high rates. By assuming that passive dye efflux is negligible at very high rates of fluid transport, the maximum rate of dye transport can be estimated. By subtracting the net dye efflux measured at lower rates of fluid perfusion from this maximum value, the passive dye efflux at each fluid perfusion rate can be calculated (Fig. 17).

In *Calliphora*, then, active dye secretion into the lumen proceeds at a high rate so that a high concentration of dye is reached in the lumen at all but high rates of fluid secretion. However, because the tubule wall is quite permeable to dye, a large proportion of transported dye diffuses passively back out of the lumen into the bathing fluid. As a result, the net amount of dye transported into the lumen depends very much on the rate of fluid secretion. In *Rhodnius*, by contrast, the tubule walls are much less permeable (Maddrell and Gardiner, 1974). As a result, much higher dye concentration gradients can be maintained across the tubule wall. This has the result that far less of the transported dye leaks back into the bathing fluid and so the net amount transported into the lumen is very little affected by the rate of fluid secretion.

Transport of the organic anion salicylate by the Malpighian tubules and gut of insects has been studied using two ion-selective microelectrode methods described above for analysis of transport of the organic cation TEA (Section 7.1; Figs. 11, 12). Salicylate is of particular interest because of its role as a signal which activates defence genes in plants in response to herbivory or pathogens. Salicylate-selective microelectrodes based on the anion exchanger tridodecylmethylammonium (TDMA) are typically >2000 times more selective toward salicylate than to Cl^-. In the first method, the concentration of salicylate in fluid droplets secreted by isolated insect Malpighian tubules set up in a Ramsay secretion assay is measured with a salicylate-selective microelectrode and flux across the entire tubule is calculated as the product of secretion rate and secreted droplet salicylate concentration (Fig. 18). In the second method, TDMA-based microelectrodes for salicylate are used in conjunction with the

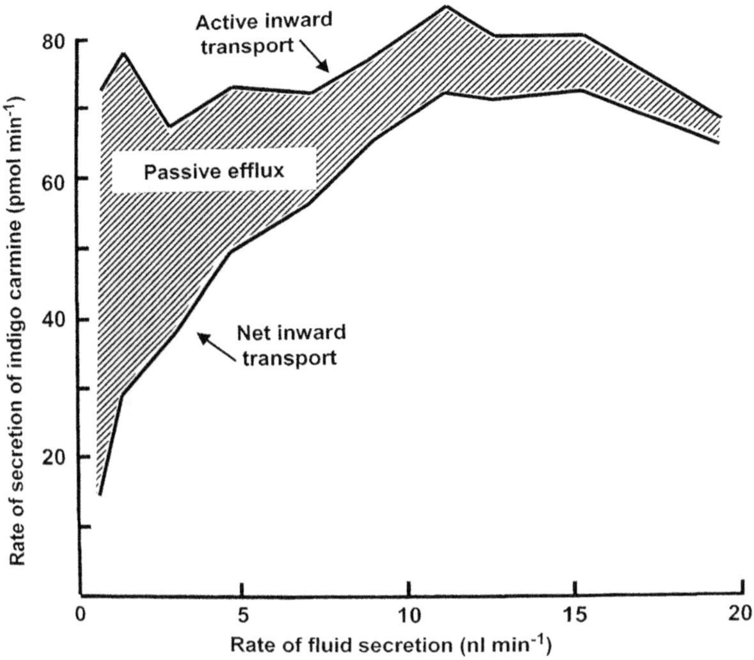

FIG. 17 Indigo carmine transport by Malpighian tubules of *Calliphora* as a function of the rate of secretion of fluid. Fluid secretion rates were varied by changing the rate of perfusion of the lumen of the cannulated tubule, as described in the text. The lower curve shows the rate of net inward dye transport actually observed. The upper curve shows the active inward transport and was constructed by adding the calculated rates of dye efflux to the lower curve, assuming the tubule wall to have a permeability to indigo carmine of $0.70\,\text{nl}\,\text{mm}^{-2}\,\text{min}^{-1}$. The stippled area between the two curves represents the extent of passive dye efflux. Redrawn from Maddrell *et al.* (1974, Fig. 14). Reprinted with permission from the Company of Biologists.

scanning ion electrode technique (SIET, also known as the self-referencing ion-selective microelectrode technique) for non-invasive spatial and temporal analysis of ion flux (Smith *et al.*, 1994). A salicylate-selective microelectrode is moved between two positions within the unstirred layer adjacent to the surface of a tissue. Salicylate transport by the tissue perturbs the salicylate concentration in the unstirred layer, and the difference in salicylate concentrations measured between the two microelectrode positions can be converted into a corresponding salicylate flux using the Fick equation.

Dose–response curves relating fluid secretion rate for *D. melanogaster* Malpighian tubules, secreted fluid salicylate concentration and salicylate flux to bathing saline salicylate concentration are shown in Fig. 18. The maximum concentration of salicylate in the secreted fluid was ~7 mM

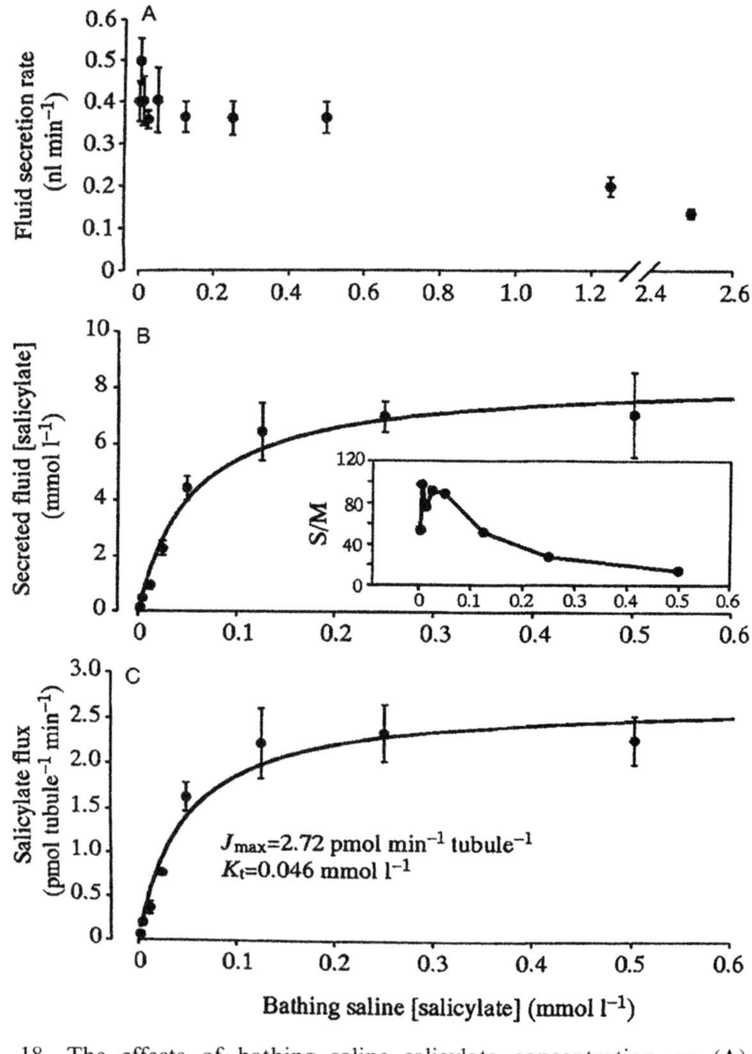

FIG. 18 The effects of bathing saline salicylate concentration on (A) fluid secretion rate, (B) secreted fluid salicylate concentration and (C) salicylate flux. Each point is the mean ± S.E.M. for 6–12 tubules. The inset in (B) shows the relationship between bathing saline salicylate concentration (mmol l^{-1}) and S/M, the ratio of salicylate concentration in the secreted fluid to that in the bathing medium. Curves in (B) and (C) were fitted to the Michaelis–Menten equation using non-linear regression analysis. J_{max} refers to the maximum rate of transport and K_t to the concentration of salicylate required to produce 50% of the maximum flux (O'Donnell and Rheault, 2005, Fig. 7). Reprinted with permission from the Company of Biologists.

(Fig. 18B). At lower bathing saline concentrations (0.0025–0.125 mM) the concentration of salicylate in the secreted fluid was elevated 50- to 100-fold (Fig. 18B, inset). Flux data were fit to the Michaelis–Menten equation to yield the values for J_{max} (2.72 pmol min^{-1}) and K_t (0.046 mM; Fig. 18C). There is good agreement between salicylate flux determined using salicylate-selective microelectrodes and that determined using liquid scintillation counting of radiolabelled salicylate (O'Donnell and Rheault, 2005). However, the microelectrode technique is faster, cheaper and provides greater resolution for small (< 1 nl) samples. However, because some competitive inhibitors of salicylate transport interfere with TDMA-based electrodes, radiolabelled salicylate is used for pharmacological characterization of transport.

Salicylate concentrations in hemolymph samples can also be measured with TDMA-based microelectrodes after injection of salicylate into the hemocoel or after insects are fed salicylate-rich diets. The rate of salicylate secretion by *Drosophila* Malpighian tubules *in vitro* is sufficient to account for the measured rate of decline of salicylate concentration in the hemolymph *in vivo*, suggesting that the Malpighian tubules play a dominant role in elimination of organic anions (O'Donnell and Rheault, 2005).

Measurements with TDMA-based microelectrodes in conjunction with the scanning ion electrode technique show that salicylate is also secreted across the midgut, ileum and rectum when the saline contains 0.5 mM salicylate (Fig. 19). When insects are fed high concentrations of salicylate in the diet there is an efflux of salicylate from gut lumen to hemolymph. SIET measurements confirmed transport of salicylate across the main and distal segments of the Malpighian tubule, but also demonstrated that there is a small efflux of salicylate (from tubule lumen to bath) across the lower Malpighian tubule. The significance of the latter finding is unclear. Perhaps organic anion levels in the hemolymph are regulated through the combined actions of the secretory and reabsorptive segments of the tubule. Taken together, the results suggest that the gut plays a role in limiting absorption of salicylate from the food, and that salicylate which does reach the hemolymph is cleared through active transport by the Malpighian tubules.

The mechanism of transepithelial salicylate transport by *Drosophila* tubules appears similar to that previously described for other organic anions (Bresler *et al.*, 1990; Linton and O'Donnell, 2000). ^{14}C-salicylate secretion by isolated tubules is competitively inhibited by other carboxylates such as probenecid, unlabelled salicylate, fluorescein and PAH acid. In common with transport of PAH and fluorescein, transepithelial transport of salicylate is strongly sodium-dependent. Transport is nearly abolished in Na^+-free saline, and is inhibited also by ouabain or K^+-free bathing saline, treatments that will reduce the Na^+ gradient across the basolateral membrane through inhibition of the Na^+/K^+-ATPase (Ruiz-Sanchez and O'Donnell, 2007a). Transepithelial secretion of

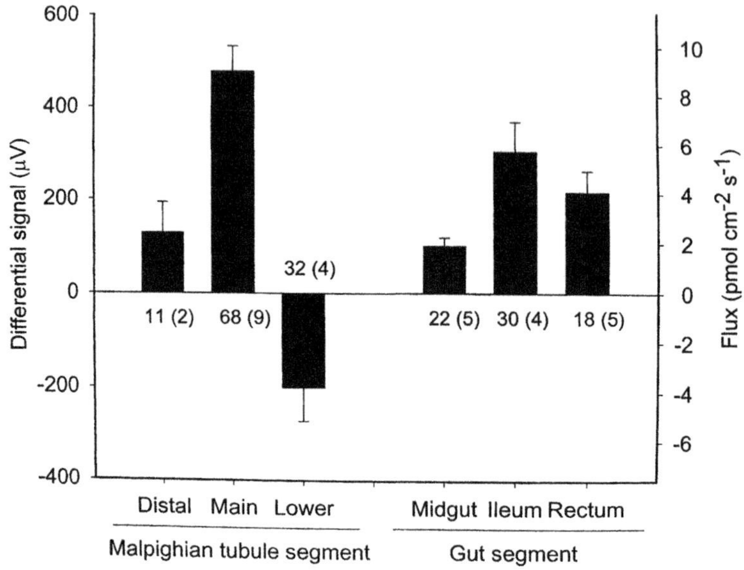

FIG. 19 Salicylate flux (right axis) and differential signal (left axis) in three segments of the *Drosophila* Malpighian tubule, and in the midgut, the ileum and the rectum. The height of each bar is the mean ±S.E.M. for the number of sites indicated at the bottom of each bar. The number of different preparations is indicated in brackets. At each site, the salicylate flux was calculated as the mean of five replicate measurements. Redrawn from O'Donnell and Rheault (2005, Fig. 5). Reprinted with permission from the Company of Biologists.

salicylate is accomplished by Malpighian tubules in at least 9 species of insect from 4 orders (Ruiz-Sanchez *et al.*, 2007).

When fluid secretion rate is altered by variations in bathing saline osmolality, 64% of the changes in salicylate transport can be explained on the basis of changes in fluid secretion rate (Ruiz-Sanchez and O'Donnell, 2007a). The increases in transepithelial salicylate transport produced by cAMP, cGMP or leucokinin are most easily explained as a consequence of the effects of these compounds on fluid secretion rate. Increases in fluid secretion rate presumably minimize passive diffusive backflux of salicylate from lumen to bath, and this effect is most apparent when the concentration of salicylate in the bathing saline is high. When the bathing saline concentration of salicylate is below the value of the Michaelis–Menten constant (K_t) the concentration in the lumen is also lower. Under these conditions the gradient for diffusive backflux is reduced and there is therefore no relationship between dye transport and fluid secretion rate (Ruiz-Sanchez and O'Donnell, 2007a).

Although renal transport of type I organic anions is Na^+-dependent in both vertebrates and insects, the mechanism of uptake of the organic anion salicylate across the basolateral membrane of the Malpighian tubules of *D. melanogaster* nonetheless differs fundamentally from mechanisms proposed for vertebrate renal tubules (Ruiz-Sanchez and O'Donnell, 2006). In vertebrate kidneys the uptake process is composed of three steps arranged in parallel. The Na^+/K^+-ATPase maintains an inwardly directed (blood to cell) Na^+ gradient which drives a Na^+-dicarboxylate cotransporter (Na^+-DC), thereby establishing an outwardly directed (cell to blood) dicarboxylate (α-keto acid) gradient. In turn, the dicarboxylate gradient is utilized by a dicarboxylate/organic anion exchanger to move the organic anion into the cell. This cascade of events indirectly links organic anion transport to metabolic energy and Na^+ transport. The activity of the organic anion uptake mechanisms in the vertebrate kidney is characteristically influenced by α-keto acids; preloading cells with these compounds (*trans*-configuration) stimulates uptake of organic anions, whereas simultaneous addition of high concentration of α-keto acids to the bathing medium (*cis*-configuration) inhibits uptake of organic anions. By contrast, preloading Malpighian tubule cells with the α-keto acid glutarate does not *trans*-stimulate salicylate or fluorescein uptake, nor is salicylate uptake *cis*-inhibited by high concentrations of various α-keto acids (α-ketoglutaric acid, glutaric acid, succinic acid and citric acid) in the saline (Linton and O'Donnell, 2000; Ruiz-Sanchez and O'Donnell, 2006). Together, these results suggest that Na^+-dependent organic anion transport by the Malpighian tubules of *D. melanogaster* is not mediated by an OAT1-like transporter similar to that of the vertebrate kidney. The absence of stimulation of salicylate uptake by acidic saline also militates against a role for a pH-dependent monocarboxylate transporter (MCT), similar to the H^+:carboxylate cotransporter of the vertebrate gut. Current models propose a Na^+:monocarboxylate transporter for salicylate uptake across the basolateral membrane of *D. melanogaster* Malpighian tubules (Fig. 6). The transport system is strongly inhibited by monocarboxylic acids with a ring structure and it is not dependent on basolateral membrane potential (Ruiz-Sanchez and O'Donnell, 2006). Inhibition of salicylate uptake by the 'classical' MCT inhibitor, α-cyano-4-hydroxycinnamic acid suggests that salicylate uptake across the basolateral membrane of the *D. melanogaster* Malpighian tubules may resemble a Na^+-dependent active monocarboxylate transport system previously reported in the brush-border membrane of rat enterocytes (Storelli *et al.*, 1980).

The monocarboxylate transporter in the basolateral membrane of the Malpighian tubule may contribute to elimination of natural or synthetic toxins from the hemolymph. For example, the plant secondary metabolite jasmonic acid, which protects plants against pathogens and phytophagous insects (Li *et al.*, 2002), would be a suitable substrate for the transport system that handles salicylate. It is also worth noting that 3-phenoxybenzoic acid, an active metabolite of the insecticide deltamethrin, is similar in structure to other substrates of the salicylate transporter (Ruiz-Sanchez and O'Donnell, 2006). One source of endogenous organic anions is through metabolism of the amino acid tryptophan. Three fluorescing materials excreta from the American cockroach, *P. americana* (L.), have been identified as kynurenic, xanthurenic and 8-hydroxyquinaldic acids (Mullins and Cochran, 1973b). Concentrations of these compounds in the excreta increase as dietary nitrogen increases, and can account for 1–3% of total nitrogen excretion. Metabolism of tryptophan through the pathway involving kynurenine and 3-hydroxykynurenine leads to the synthesis of ommochrome eye pigments. In addition, kynurenine can be converted to a number of intermediates some of which are excreted.

8.3.1 *Regulation of organic anion transport by Malpighian tubules*

Transport of organic anions such as PAH and amaranth by the Malpighian tubules of *R. prolixus* varies with the insect's physiology (Maddrell and Gardiner, 1975). PAH transport declines gradually as starved insects age, but increases dramatically by more than 20-fold within 1–2 days after a blood-meal (Fig. 9). A similar pattern is seen for urate transport, as described above (Fig. 9). Thereafter the rate declines with a half-time of about 10–15 days. Ingestion of a meal of glucose-free Ringer's solution does not result in an increase in the rate of PAH transport, which suggests that the normal induction of transport is not controlled by a hormone released in response to the same stimuli that elicit release of the diuretic and prothoracicotropic hormones. Ingestion of blood is not the only effective stimulus for induction of PAH excretion. Meals containing protein in the form of blood plasma alone, a suspension of red blood cells in saline, serum albumin, casein or milk also result in an increase in PAH transport several days later. However, there is no increase in rates of excretion of organic anions after a single injection of 0.1 μmol of PAH into the hemolymph. Ingestion of a protein-rich meal is thus sufficient to induce an accelerated transport of organic anions and this induction is not controlled by a hormone released in response to abdominal distension. Rather, induction depends instead on the sustained presence in the hemolymph of some product of digestion of the meal. As noted by Maddrell and Gardiner (1975), it is unclear why the transport of organic anions such as PAH should be regulated in this way, whereas the transport of the alkaloids nicotine and atropine is maintained irrespective of the insect's nutritional state.

Changes in organic anion transport in response to dietary loading with organic anions have also been documented in third instar larvae of *D. melanogaster* (Ruiz-Sanchez and O'Donnell, 2007b). Larvae chronically exposed to dietary salicylate show lower levels of salicylate in the hemolymph and increased elimination of salicylate after feeding on salicylate-enriched diet. Exposure to dietary salicylate also leads to a fivefold increase in salicylate excretion by isolated Malpighian tubules. This increase is accompanied by an increase of more than threefold in Malpighian tubule fluid secretion rate. The increases in fluid and salicylate secretion do not appear to reflect increases in the basal levels of the second messengers cAMP and Ca^{2+} which increase apical proton pump activity and transepithelial Cl^- permeability, respectively. Instead, increases in the unstimulated level of fluid secretion in larvae fed a salicylate-rich diet may indicate increased numbers of transporters (e.g. H^+-ATPase) in the tubule cells.

An important consequence of the increase in fluid secretion rate is that the concentration of salicylate in the tubule lumen is maintained at relatively low levels. Diffusive backflux of salicylate from the tubule lumen to the hemolymph is thus minimized. Transepithelial salicylate transport increases with fluid secretion rate irrespective of whether fluid secretion is increased by intracellular second messengers or changes in bathing saline osmolality (Ruiz-Sanchez and O'Donnell, 2007a). Taken together the results indicate that exposure to salicylate in the diet leads to alterations in renal function that enhance clearance of the toxin from the insect's hemolymph. Moreover, the increase in fluid secretion rate will enhance the excretion of other toxins.

8.3.2 *Excretion of insecticides and their metabolites*

One component of resistance to the organophosphate insecticide leptophos in the Egyptian cotton leafworm *Spodoptera littoralis* (Boisd.) is more rapid excretion of both the unchanged insecticide and its metabolite leptophos phenol (Ben-Aziz *et al.*, 1976; Riskallah and Esaac, 1982). The half-time ($t_{1/2}$) for disappearance of leptophos in larvae of a resistant strain ($t_{1/2} = 2.5\,h$) is > 8 times shorter than in a larvae of a susceptible strain ($t_{1/2} = 21.5\,hr$). Excretion of leptophos metabolites is also more rapid in resistant than in susceptible strains. The major degradation product is the free phenol and its amount is relatively higher in resistant (17–22%) than in susceptible (5–15%) larvae. The rate of excretion of the phenol is fourfold higher in resistant than in susceptible larvae. Esterase activity is consistently higher in resistant strains of the leafworm, enabling them to degrade leptophos at a higher rate than in susceptible strains.

Control of organophosphate-resistant strains of the diamondback moth *Plutella xylostella* necessitates the use of compounds such as malathion at doses several thousand fold higher than those required elimination of susceptible strains. Resistance to malathion involves three changes in the physiology of the insect. Although malathion penetrates the cuticle of insects from resistant strains more rapidly, it is also metabolized and excreted more rapidly, relative to insects from susceptible strains

(Doichuanngam and Thornhill, 1992). As for *Sp. littoralis*, metabolic degradation of a toxin is thus coupled to enhanced excretion of the degradation products.

Enhanced excretion of insecticides has also been described for other species. Resistant larvae of the Egyptian cotton leafworm larvae also excrete carbaryl (Hanna and Atallah, 1971). Similarly, resistant houseflies excrete the parathion metabolite paraoxon (Plapp *et al.*, 1961) and resistant larvae of the tobacco budworm houseflies excrete large amounts injected endrin (Polles and Vinson, 1972). Resistant larvae in each case can tolerate higher doses of insecticidal chemicals by the simple expedient of excreting the toxic chemical intact.

Although most of the studies of excretion of insecticide metabolites have been based on *in vivo* experiments and collection of the excreta for analysis, recent *in vitro* studies have suggested a specific role for the Malpighian tubules. Analysis of the transport of fluorescein by the Malpighian tubules of the cricket *Ac. domesticus* suggests that organic anion transporters in the tubules play a role in excretion of some malathion metabolites (Neufeld *et al.*, 2005). Esterase activity converts malathion into malathion monocarboxylic acid (MMA) which competitively inhibits transport of the organic anion fluorescein by the tubules. Inhibition of accumulation of MMA by the tubules by probenecid, an archetypal organic anion transport inhibitor, was verified in this study by high performance liquid chromatography (HPLC). Transport of MMA by the tubules suggests that transport processes may confer resistance to xenobiotics in insects (Neufeld *et al.*, 2005).

8.4 TRANSPORT OF TYPE II ORGANIC ANIONS BY MULTIDRUG RESISTANCE-ASSOCIATED PROTEINS

The MRPs are another group of ATP-binding cassette (ABC) transporters implicated in excretion of toxins by insects. As for P-gps, MRPs are known for their role in conferring resistance to antineoplastic agents in humans. Whereas P-gp substrates are primarily cationic or neutral compounds, MRP substrates are more commonly anions, although some neutral compounds, cations and even heavy metals such as arsenic are also transported by members of this family (Leslie *et al.*, 2005; Deeley *et al.*, 2006).

MRPs in insects were first identified in the Malpighian tubules of the cockroach *P. americana* and the cricket *Ac. domesticus* by Karnaky *et al.* (2000, 2001, 2003). Malpighian tubules of both species transport the fluorescent MRP2 substrate Texas Red (sulforhodamine 101) through the cells and into the lumen. Analysis of isolated Malpighian tubules using confocal microscopy showed that the concentration of the fluorochrome in the lumen is dependent upon metabolism and is inhibited by the non-fluorescent MRP2 substrate chlorodinitrobenzene, which does not inhibit lumenal accumulation of the smaller, more hydrophilic organic anion chlorophenol red. Transport is unaffected by addition of the organic anion transporter substrate PAH acid, the organic cation transporter substrate TEA chloride, or by the P-gp inhibitor verapamil (Karnaky *et al.*, 2001). Luminal cells are immunopositive for an antibody to a sequence of rat MRP2, further suggesting that a MRP2-like transporter is present in Malpighian tubules.

Transepithelial flux of the MRP2 substrate Texas Red across the Malpighian tubules of the fruit fly, *D. melanogaster* and the cricket, *Te. commodus* has been measured using confocal laser scanning microscopy (Leader and O'Donnell, 2005). Texas Red accumulates in the cells and lumen of cricket tubules at concentrations up to 20 and 40 times, respectively, those in the lumen. Transepithelial flux of Texas Red is calculated by multiplying the dye concentration (μM), as measured by CLSM of secreted fluid samples collected in rectangular glass capillaries, by the fluid secretion rate (nl min^{-1}) measured in the Ramsay assay. Estimates based on the *in vitro* fluxes suggest that *Drosophila* tubules could clear the hemolymph of 20 μM Texas Red within 6 min. Transport of Texas Red is inhibited by the MRP2 inhibitors MK-571 and probenecid. However, Texas Red transport is not reduced in Na^+-free saline, suggesting that probenecid is not acting on the Na^+-dependent process implicated in transepithelial transport of smaller organic anions such as fluorescein (Linton and O'Donnell, 2000) and salicylate (Ruiz-Sanchez and O'Donnell, 2006). Analysis of concentration-response curves for secretion of Texas Red by Malpighian tubules indicates that the transport mechanism is characterized by higher affinity (i.e. lower K_t) but lower capacity (J_{max}) than the transporters utilized for Na^+-dependent-transport of fluorescein and salicylate. The value of K_t for Texas Red (7.1 μM) is fivefold lower than that for salicylate (36 μM), whereas the J_{max} for Texas Red (118 fmol min^{-1} $tubule^{-1}$) is 23-fold lower than that for salicylate (2.72 pmol min^{-1} $tubule^{-1}$).

The effects of stimulants of fluid secretion on net transepithelial transport of P-gp and MRP2 substrates by *Drosophila* Malpighian tubules have been examined by O'Donnell and Leader (2006). Stimulation of fluid secretion might alter the transport of an organic ion by altering the transepithelial potential. Both cAMP and cGMP produce a lumen positive potential in *Drosophila* tubules, thus tending to favour secretion of anions such as Texas Red. Tyramine augments fluid secretion through a Ca^{2+}-dependent increase in transepithelial Cl^- permeability, reducing the magnitude of the lumen positive transepithelial potential. Tyramine might thus enhance the transport of type II organic cations such as daunorubicin by reducing the opposing electrical gradient. Increases in the levels of intracellular second messengers in response to diuretic factors might also increase the levels of protein kinases (e.g. protein kinase C). Studies of vertebrate tissues have demonstrated the modulation of P-gps and MRP2 by protein kinase C, raising the possibility of some form of MRP2 modulation by protein kinases in insect Malpighian tubules.

Alternatively, stimulants of fluid secretion may alter transepithelial flux of MRP2 substrates by indirect means. An increase in fluid secretion rate will tend to dilute the MRP2 substrate in the lumen, thereby reducing the tendency for diffusive backflux of the dye from the lumen to the bathing medium. Increases in fluid secretion rate have the greatest effect on solutes that are actively transported (i.e. when the solute concentration in the lumen is much greater than that in the bathing medium) and to which the tubule is relatively permeable (Maddrell *et al.*, 1974). For the MRP2 substrate Texas Red, dye transport is greatly affected by fluid secretion rate (O'Donnell and Leader, 2006). When the rate of fluid secretion is stimulated by cAMP or tyramine, transport of Texas Red is increased, and these effects are more pronounced when the bathing saline concentration of the dye is near or above the concentration required for half-maximal transport (K_t). Regression analysis indicates that 57–80% of the change in dye flux is attributable to changes in fluid secretion rate produced by the stimulants. Similarly, when the osmolality of saline bathing cAMP-stimulated tubules is increased 22% by addition of sucrose, the reduction in fluid secretion rate accounts for 88% of the reduction in Texas Red flux. Taken together, these results suggest that increases in fluid secretion rate minimize diffusive backflux, and that stimulants of fluid secretion thus increase transepithelial fluxes of Texas Red indirectly. A similar relationship between fluid secretion rate and solute flux has been described for Na^+-dependent transport of the type I organic anion salicylate (Ruiz-Sanchez and O'Donnell, 2007a) in Section 7.3.1.

8.4.1 *Metabolism and excretion of acetylchromenes*

Precocenes, natural 2,2-dimethyl-chromene constituents of the plant *Ageratum houstoniunum*, destroy the corpora allata (the source of juvenile hormone) in sensitive insect species and thereby cause precocious moulting. However, precocenes are atypical of the >30 more widely distributed plant chromenes in which an acetyl group is linked to the aromatic ring. The insecticidal compound Encecalin (from *Encelia* spp., Asteraceae) is one of the commonest acetylchromenes. Only 20–40% of the applied dose of encecalin and demethylencecalin (DME) is recovered in the frass of treated grasshoppers (*Me. sanguinipes*) after 48 h, almost all of which is excreted within 24 h of administration. Both parent compounds and metabolites of acetylchromenes and precocene II are excreted and the metabolites are the result of phase 1 reactions (Isman *et al.*, 1987). Other studies of precocene II metabolism in insects report that the major route of precocene metabolism in several species of insects is via demethylation followed by rapid glucosylation. Much of the radiolabel in hemolymph of the African armyworm *Spodoptera exempta* following treatment with labelled precocene II is also in the form of conjugates. Acetylchromene metabolites may occur as conjugates in the hemolymph and may be hydrolyzed subsequently with recovery of the glucose (or other conjugating group) during excretion by the Malpighian tubules (Isman *et al.*, 1987). Given the metabolism in phase I reactions and conjugation of acetylchromene metabolites, it will be of interest to determine whether P-gps or MRPs in the Malpighian tubules are involved in excretion of these compounds. In other systems, P-gps often act on the product of phase I reactions, such as those mediated by P450 enzymes. Similarly, there is evidence that MRPs act on conjugated compounds produced through the actions of glutathione-*S*-transferases, as described in Section 8.5, below.

8.4.2 *Excretion of sesquiterpene lactones and their metabolites*

It will also be of interest in future studies to examine the role of transporters such as P-gps and MRP2 in the excretion of a number of plant-derived insecticides such as the sesquiterpene lactones. The leaves of many trees in the family Magnoliaceae, such as the tulip tree (*Liriodendron tulipifera* L.) and sweet bay (*Magnolia virginiana* L.), contain a variety of toxic constituents, including sesquiterpene lactones such as parthenolide. Sesquiterpene lactones are feeding deterrents and their toxicity is due to the presence of various electrophilic groups, such as the α-methylene-γ-lactone. The mechanism of toxicity is relatively non-specific because these electrophilic groups react with electron-rich atoms such as sulfur and oxygen in proteins and nucleic acids. Parthenolide also causes oxidative stress in cells, leading to gut lesions and apoptosis, and this stress may be exacerbated by a concomitant depletion of the antioxidant molecule glutathione.

Caterpillars of the tiger swallowtail butterfly *Papilio glaucus* consume large amounts of parthenolide when feeding on in the leaves of host plants such as tulip trees and sweet bay. They avoid the toxic effects of diets which may contain as much as 0.5% parthenolide by excreting intact sesquiterpene lactones and sesquiterpene lactone metabolites (Frankfater *et al.*, 2005). Excretion of unaltered parthenolide from the diet indicates that either there is a barrier to absorption of allelochemicals across the gut or that there is absorption into the hemolymph followed by highly efficient excretion by the Malpighian tubules. Parthenolide is modified in the gut by reduction of an exocyclic methylene group, possibly by an (NADPH)-dependent reductase, such as an α,β-ketoalkene double bond reductase, and also by addition of a hydroxyl group, presumably in a reaction catalyzed by a cytochrome P450 enzyme. These reactions may be accomplished either by gut microbes or by enzymes present in the midgut or Malpighian tubules. Sesquiterpenes are transported by P-gps in mammalian cells, and the actions of P450 enzymes may render them as more suitable substrates (Barthomeuf *et al.*, 2006). Alternatively, the increase in polarity of the metabolite as a consequence of the addition of the hydroxyl group may promote subsequent conjugation to sugars or glutathione and thereby lead to enhanced excretion by transporters such as the MRP2. It is worth noting in this context that high levels of MRP2-like gene expression are found in the gut of another lepidopteran, the cabbage looper *Tr. ni* (R. Labbe and D. C Donly, personal communication).

8.4.3 *Secretion of cyclic nucleotides by Malpighian tubules*

Organic anion transport has also been implicated in the secretion of cyclic nucleotides by insect Malpighian tubules. Cyclic AMP is secreted by isolated tubules of *R. prolixus* in response to stimulation with 5-HT and the compound is also excreted in the urine during diuresis *in vivo* (Montoreano *et al.*, 1990). The cells of the Malpighian tubules of *Drosophila* also take up cAMP and cGMP from the bath (Riegel *et al.*, 1998) and uptake is inhibited by the organic anion transport inhibitor probenecid and by other organic anions such as PAH and phenolsulfonthalein, but not by ouabain or the weak base quinacrine (Riegel *et al.*, 1999). Cyclic GMP appears in the fluid secreted by *Drosophila* tubules (Day *et al.*, 2006), indicating that there is also transport of it and presumably other cyclic nucleotides from cell to lumen. The uptake of cyclic nucleotides across the basolateral membrane may be mediated by the same Na^+-dependent mechanism which transports PAH and salicylate. The mechanism of transport across the apical membrane into the lumen is unknown, although it is worth noting that members of the MRP family transport cyclic nucleotides in mammalian cells (Deeley *et al.*, 2006).

8.5 INTERACTION OF EXCRETORY MECHANISMS WITH PHASES I AND II DETOXIFICATION MECHANISMS

Detoxification through phases I and II mechanisms is commonly implicated in insecticide resistance. Elevated levels of the phase II enzyme glutathione-*S*-transferase are associated with resistance to all major classes of insecticides (Ranson and Hemingway, 2005). Similarly, the P450 monoxygenases are phase I enzymes which are well known for their roles in metabolism of xenobiotics, including natural and synthetic insecticides, to less toxic metabolites. However, as noted by Feyereisen (1999), "P450 enzymes will metabolize xenobiotics irrespective of the eventual fate of the metabolites, and there are many cases where P450 enzymes act as anything but detoxification enzymes." Indeed, P450 chemistry has been exploited to produce bioactivated insecticides. Metabolism of pesticides to more toxic compounds through the action of P450 enzymes is the basis for the efficiency of the phosphorothioate insecticides (conversion of P = S to P = O) and the cyclodiene insecticides (aldrin and heptachlor to their epoxides).

There are several reasons to suspect that phase I and phase II pathways may interact with phase III elimination pathways in insects. Several cytochrome P450 genes and glutathione-*S*-transferase genes are enriched in the Malpighian tubules of *Drosophila* (Dow and Davies, 2006). It is also well known that P-gps and P450 enzymes share many substrates (Bard, 2000; Abu-Qare *et al.*, 2003). A moderately hydrophobic xenobiotic may therefore be transported out of cells unmodified or after hydroxylation via P450 enzymes (Fig. 20). Similarly, xenobiotic resistance mediated by GST may require removal of the glutathione conjugate via MRP2 (e.g. Smitherman *et al.*, 2004). Some xenobiotics may therefore be transported

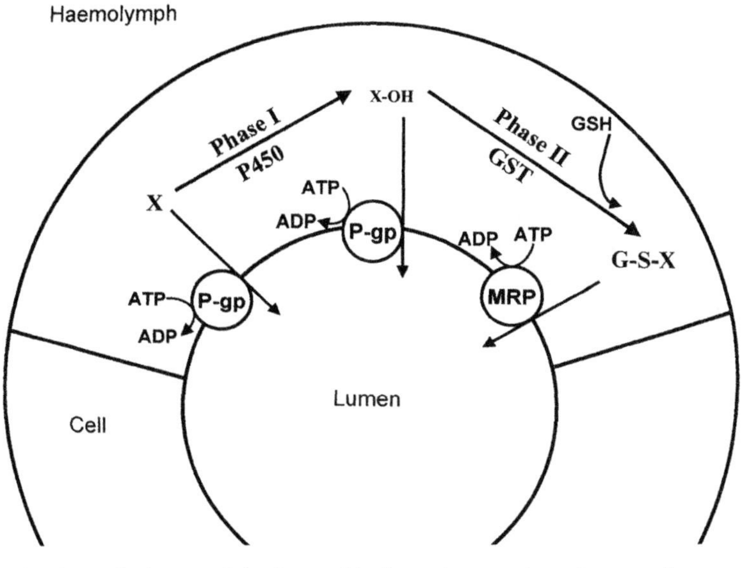

FIG. 20 Speculative model of xenobiotic resistance based on active transport through apical P-glycoprotein (P-gp) and multidrug resistance-associated protein (MRP), the Phase I detoxification enzymes, cytochromes P450 (P450) and the phase II detoxification enzyme glutathione-*S*-transferase (GST). It is suggested that the transporters reside in the apical membrane of the Malpighian tubule and gut. A moderately hydrophobic toxin (X) may be transported into the lumen by P-gp. Alternatively, the compound may be modified by P450 enzymes and either transported into the lumen by P-gp, or further modified by conjugation with glutathione (GSH) through the actions of GST and then transported into the lumen by MRP. Redrawn and modified from Fig. 3 of Bard (2000). Reprinted with permission of Elsevier Science.

by MRP2 either unmodified or subsequent to both hydroxylation by P450 enzymes and conjugation by GST (Fig. 20).

The esterases are another group of enzymes that play important roles in insecticide resistance (Hemingway and Ranson, 2000). Many examples of resistance to organophosphorous insecticides are based on increased detoxification of the insecticide by esterases. In the mosquito *Culex pipiens* resistance involves overproduction of esterases which have a high binding affinity for insecticides but a low rate of hydrolysis. As a result, the insecticide is sequestered and unable to reach the target enzyme acetylcholinesterase. Overproduction of the esterases involves either increased gene expression or, more commonly, gene amplification. A common feature of resistant strains of *Culex* is overproduction of esterases in the alimentary canal and the Malpighian tubules (Pasteur *et al.*, 2001). Esterases in the Malpighian tubules are aptly located both to aid detoxification of the hemolymph and to facilitate subsequent excretion of the products of esterase activity. Esterases in the alimentary canal presumably limit transfer of active insecticide from the gut to the hemolymph, from where it can gain access to the target enzymes of the nervous system.

Given the common finding that phase I and phase II enzymes are induced by exposure to appropriate substrates, it will be of interest to see if there is an associated up regulation of phase III elimination pathways. The effects of the plant-produced signal salicylate are of interest in this context. Salicylate activates plant defence genes after herbivory or pathogenic attack. The corn earworm can 'eavesdrop' on this signal, so that by the time salicylate activates production of toxic allelochemicals in the corn plants, the insect's own P450 enzymes have been activated (Li *et al.*, 2002). However, chronic exposure to salicylate is also known to increase fluid secretion rate and salicylate secretion by isolated Malpighian tubules of *D. melanogaster* (Ruiz-Sanchez and O'Donnell, 2007b). As a result, salicylate is more rapidly cleared from the hemolymph. Given that phase I and phase II pathways modify the chemical properties of molecules to make them more suitable for elimination by transporters such as P-gps and/or MRPs, it would seem beneficial for insects to increase the capacity of their phase III pathways during exposure to xenobiotics. Exploration of the links between the detoxification (phase I and II) and elimination pathways (phase III) may thus lead to new insights into understanding insecticide resistance in pest species. In this context, it is worth noting the study by Yang *et al.* (2007) of insecticide metabolism in *Drosophila*. This study shows that cytochrome P450 and GST loci implicated in insecticide resistance are most abundantly expressed in the Malpighian tubule. The importance of the tubule in detoxification of insecticides is reinforced by the finding that flies become more susceptible to DDT intoxication if RNAi against just one P450, cyp6g1, is expressed in only the tubule cells of transgenic *Drosophila*. This implies that the tubule can be the limiting tissue in insecticide metabolism.

9 Future directions

9.1 THE IMPACT OF GENOME SEQUENCING PROJECTS

The sequencing of the *Drosophila* genome has had a dramatic impact upon our understanding of ion transport mechanisms (Section 3.2), control of ion transport by extracellular and intracellular messengers and the great diversity of solute transporters in the *Drosophila* tubule (Wang *et al.*, 2004). Sequencing of the genomes for other insect species will allow expression of the genes for specific ion channels, pumps, cotransporters and exchangers to be studied in the absence of other transporters. By 2007, whole-genome sequencing projects have been completed for 12 insect lineages (Zdobnov and Bork, 2007): the honeybee (*Apis mellifera*) representing order Hymenoptera, the red flour beetle (*Tribolium castaneum*) from the Coleoptera, the silkworm moth (*B. mori*) from the Lepidoptera and two mosquitoes (*An. gambiae* and *Ae. aegypti*) and seven fly species (*D. melanogaster*, *Drosophila erecta*, *Drosophila ananassae*, *Drosophila pseudoobscura*, *Drosophila mojavensis*, *Drosophila virilis* and *Drosophila grimshawi*) from the Diptera. Many important advances in our understanding of insect excretory mechanisms will be based on heterologous expression of the genes for specific transporters from the gut or Malpighian tubules in cell lines or *Xenopus* oocytes, so that the transporter may be characterized in isolation from other transporters present in the whole epithelium. Studies of organic ion transporters using heterologous expression may be particularly important because Malpighian tubule cells, like other renal epithelia, possess multiple transport systems with overlapping substrate specificities. An alternative approach is the use of gene-knockouts or the selective reduction of the expression of an individual gene through the use of RNAi techniques (e.g. Torrie *et al.*, 2004). The use of tissue-specific DNA microarrays, such as that developed for the *D. melanogaster* Malpighian tubule (Wang *et al.*, 2004), will be of great use in looking at the functions of the excretory system in responding to stresses such as high levels of toxins in the diet, and the interactions between membrane transporters and phase I and phase II detoxification pathways.

9.2 CHEMICAL ECOLOGY AND INSECT PHYSIOLOGY

It also appears likely that our understanding of insect excretory system will be aided by advances in chemical ecology. Improvements in separation and identification of organic compounds in the diet or excreta of insects will aid the design of *in vivo* and *in vitro* experiments designed to reveal the sites and mechanisms of excretion. For example, compounds identified from the frass of insects feeding on particular plants may be applied to isolated Malpighian tubules *in vitro* with a view to determining if the compounds are secreted by the tubules, and, if so, which transporters are involved. More detailed understanding of insect excretory mechanisms through these types of experiments will aid the development of novel and environmentally-benign control measures for agricultural pests and disease vectors.

9.3 ION CHANNELS AND INSECT EXCRETORY MECHANISMS

Further application of electrophysiological techniques, including patch clamping, double-barrelled intracellular ion-selective microelectrodes and the scanning ion-selective electrode technique, will extend our understanding of the mechanisms of ion transport in insect excretory tissues. Patch clamp studies to date have examined the role of Cl^- channels in the tubules of dipterans (O'Donnell *et al.*, 1998; O'Connor and Beyenbach, 2001). These studies have dealt primarily with channels in the apical membrane, which can be exposed by microdissection techniques (Section 3.2.4). Patch clamping of the basolateral membrane requires movement of the basement membrane. A clever mechanical technique for removal of the basement membrane from the tubules of *Onymacris rugatipennis* has been described by Nicolson *et al.* (1991) and Nicolson and Isaacson (1990). The basement membrane of the tubules of *R. prolixus* and *D. melanogaster* has been removed enzymatically using elastase (Levinson and Bradley, 1984). Single channel currents have been successfully recorded form tubules of both species treated with elastase (O'Donnell, unpublished observations), but the nature and characteristics of the channels have yet to be determined.

There is a paucity of information on the characteristics and regulation of K^+ channels in the basement membrane, in spite of the importance of K^+ channels to models of transepithelial ion secretion for Malpighian tubules of several species. An important study by Wu and Beyenbach (2003) has demonstrated the control of basolateral K^+ channels and also the apical V-type H^+-ATPase by ATP. When intracellular ATP levels decline to 10% of the control level, the V-ATPase is inhibited and K^+ movement from cell to lumen is prevented. Basolateral K^+ channels close at the same time, preventing exchange of K^+ across the basolateral membrane. These changes facilitate intracellular K^+ homeostasis when metabolism is inhibited, and allow the epithelium to re-initiate transepithelial K^+ transport when ATP levels are restored.

Intracellular recording techniques and secretion assays suggest the presence of ATP-regulated potassium channels (K_{ATP}) in the basolateral membranes of the tubules of the mealworm *Te. molitor* and the fruit fly *D. melanogaster* (Wiehart *et al.*, 2003; Evans *et al.*, 2005). The K_{ATP} channels are in addition to the Ba^{2+}-sensitive K^+ channels in the membrane (Wiehart *et al.*, 2003). Patch clamp studies, in particular single channel recordings, will be of great use in determining the types and of K^+ channels present and their contributions to parameters such as the basolateral membrane potential and net transepithelial secretion as well as homeostatic functions such as cell volume regulation. Similarly, patch clamping will be of use in examining the characteristics and regulation of the amiloride-sensitive and cAMP-induced Na^+ conductance in the principal cells of *Ae. aegypti* Malpighian tubule (Sawyer and Beyenbach, 1985; Beyenbach and Masia, 2002) as described in Section 3.2.3, and in determining the extent of similarity between this conductance and that associated with the epithelial Na^+ channel (EnaC) in vertebrate cells. Ion channels for K^+ and Cl^- may also be involved in the reabsorption of these ions by the proximal (lower) Malpighian tubule of *R. prolixus* (Haley and O'Donnell, 1997; Haley *et al.*, 1997) and *D. melanogaster* (O'Donnell and Maddrell, 1995), as described in Section 4.1. In addition to helping to understand the mechanisms of transepithelial ion secretion and reabsorption and the regulation of channels by second messengers and kinases, patch clamp studies will be of use in identification of channel blockers which may aid development of novel control measures for pest species of insect.

9.4 DIURETIC AND ANTIDIURETIC AGENTS AS CANDIDATES FOR DEVELOPMENT OF NOVEL INSECTICIDES

Control of Malpighian tubules and the insect hindgut by peptides has been discussed in Sections 3 and 4, respectively. One reason for the intense interest in studying insect peptides is the possibility of developing species-specific insecticides. Unfortunately, peptides such as insect kinins were long considered unsuitable as pest control agents because they are susceptible to both exo- and endopeptidases in the hemolymph and gut of the insect. However, when critical residues in the C-terminal pentapeptide core of insect kinins are replaced with β-amino acids or their β-homo-amino acid counterparts, potent diuretic activity is retained and the compounds are resistant to the endopeptidases which deactivate the natural insect kinins (Zubrzak *et al.*, 2007). Similarly, replacement of the N-terminal lysine of a kinin with aminohexanoic acid provides enhanced resistance to aminopeptidase attack and retains biological activity (Nachman *et al.*, 2002). Peptidase-resistant insect kinin analogs inhibit weight gain in larvae of the agriculturally destructive corn earworm moth (*ibid*), suggesting that peptidase resistant analogs of the insect kinins, may aid development of new, environmentally benign control measures for pest species.

The strategies for making kinins insensitive to breakdown by peptidases may in future be applied to other insect peptides, such as those controlling the hindgut (Section 4.2).

Acknowledgements

Research in the author's laboratory is supported by the Natural Sciences and Engineering Research Council of Canada. The author is grateful to two anonymous referees for their comments, and also to Andrew and Elizabeth Forester for their hospitality during the writing of parts of this chapter.

References

Abu-Qare, A. W., Elmasry, E. and Abou-Donia, M. B. (2003). A role for P-glycoprotein in environmental toxicology. *J. Toxicol. Environ. Health B. Crit. Rev.* **6**, 279–288.

Ainsworth, C., Wan, S. and Skaer, H. (2000). Coordinating cell fate and morphogenesis in *Drosophila* renal tubules. *Philos. Trans. R. Soc. Lond. B Biol. Sci.* **355**, 931–937.

Al-Fifi, Z. I. A., Anstee, J. H. and Bowler, K. (1998). The action of inhibitors of protein kinases on fluid and ion secretion by Malpighian tubules of *Locusta migratoria*, L. *J. Insect Physiol.* **44**, 973–980.

Allan, A. K., Du, J., Davies, S. A. and Dow, J. A. (2005). Genome-wide survey of V-ATPase genes in *Drosophila* reveals a conserved renal phenotype for lethal alleles. *Physiol. Genomics* **22**, 128–138.

Aneshansley, D. J., Marler, C. E. and Beyenbach, K. W. (1988). Transepithelial voltage measurements in isolated Malpighian tubules of *Aedes aegypti*. *J. Insect Physiol.* **35**, 41–52.

Anstee, J. H., Bell, D. M. and Fathpour, H. (1979). Fluid and cation secretion by the Malpighian tubules of *Locusta*. *J. Insect Physiol.* **25**, 373–380.

Audsley, N., Coast, G. M. and Schooley, D. A. (1993). The effect of *Manduca sexta* diuretic hormone on fluid transport by the Malpighian tubules and cryptonephric complex of *Manduca sexta*. *J. Exp. Biol.* **178**, 231–243.

Bain, L. J. and LeBlanc, G. A. (1996). Interaction of structurally diverse pesticides with the human MDR1 gene product P-glycoprotein. *Toxicol. Appl. Pharmacol.* **141**, 288–298.

Balshin, M. and Phillips, J. E. (1971). Active absorption of amino acids in the rectum of the desert locust (*Schistocerca gregaria*). *Nature* **233**, 53–55.

Bard, S. M. (2000). Multixenobiotic resistance as a cellular defense mechanism in aquatic organisms. *Aquat. Toxicol.* **48**, 357–389.

Barthomeuf, C., Demeule, M., Grassi, J., Saidkhodjaev, A. and Beliveau, R. (2006). Conferone from *Ferula schtschurowskiana* enhances vinblastine cytotoxicity in MDCK-MDR1 cells by competitively inhibiting P-glycoprotein transport. *Planta Med.* **72**, 634–639.

Bednarczyk, D., Mash, E. A., Aavula, B. R. and Wright, S. H. (2000). NBD-TMA: a novel fluorescent substrate of the peritubular organic cation transporter of renal proximal tubules. *Pflug. Archiv.* **440**, 184–192.

Begun, D. J. and Whitley, P. (2000). Genetics of alpha-amanitin resistance in a natural population of *Drosophila melanogaster*. *Heredity* **85**, 184–190.

Ben-Aziz, A., Meisner, J., Aharonson, N. and Ascher, K. R. S. (1976). Excretion of unchanged organophosphorus compounds after their consumption by larvae of the Egyptian cotton leafworm, *Spodoptera littoralis* (Boisd.). *Pestic. Biochem. Physiol.* **6**, 58–64.

Beyenbach, K. W. (1990). Transport of magnesium across biological membranes. *Magnes. Trace Elem.* **9**, 233–254.

Beyenbach, K. W. (2003). Transport mechanisms of diuresis in Malpighian tubules of insects. *J. Exp. Biol.* **206**, 3845–3856.

Beyenbach, K. W. and Masia, R. (2002). Membrane conductances of principal cells in Malpighian tubules of *Aedes aegypti*. *J. Insect Physiol.* **48**, 375–386.

Beyenbach, K. W. and Wieczorek, H. (2006). The V-type H^+-ATPase: molecular structure and function, physiological roles and regulation. *J. Exp. Biol.* **209**, 577–589.

Bijelic, G. and O'Donnell, M. J. (2005). Diuretic factors and second messengers stimulate secretion of the organic cation TEA by the Malpighian tubules of *Drosophila melanogaster*. *J Insect Physiol.* **51**, 267–275.

Bijelic, G., Kim, N. and O'Donnell, M. J. (2005). Effects of dietary or injected organic cations on larval *Drosophila melanogaster*: mortality and elimination of tetraethylammonium from the haemolymph. *Archiv. Insect. Biochem. Physiol.* **60**, 93–103.

Bradley, T. J. (1985). Comprehensive insect physiology, biochemistry and pharmacology. In: *The Excretory System: Structure and Function*, vol. 4 (eds Kerkut, G. A. and Gilbert, L. I.), pp. 421–465. London and New York: Pergamon.

Bradley, T. J. (1987). Physiology of osmoregulation in mosquitoes. *Ann. Rev. Entomol.* **32**, 439–462.

Bradley, T. J. and Phillips, J. E. (1977). The location and mechanism of hyperosmotic fluid secretion in the rectum of the saline-water mosquito larvae *Aedes taeniorhynchus*. *J. Exp. Biol.* **66**, 111–126.

Bradley, T. J. and Satir, P. (1979). Insect axopods. *J. Cell Sci.* **35**, 165–175.

Braeckman, B., Smagghe, G., Brutsaert, N., Cornelis, R. and Raes, H. (1999). Cadmium uptake and defense mechanism in insect cells. *Environ. Res.* **80**, 231–243.

Bresler, V. M., Belyaeva, E. A. and Mozhayeva, M. G. (1990). A comparative study on the system of active transport of organic acids in Malpighian tubules of insects. *J. Insect Physiol.* **36**, 259–270.

Buchanan, B. B., Gruissem, W. and Jones, R. L. (2000). *Biochemistry and Molecular Biology of Plants*. Rockville, MD: American Society of Plant Physiologists.

Buchwalter, D. B. and Luoma, S. N. (2005). Differences in dissolved cadmium and zinc uptake among stream insects: mechanistic explanations. *Environ. Sci. Technol.* **39**, 498–504.

Buckner, J. S. (1982). Hormonal control of uric acid storage in the fat body during last-larval instar of *Manduca sexta*. *J. Insect Physiol.* **28**, 987–993.

Buckner, J. S. and Newman, S. M. (1990). Uric acid storage in the epidermal cells of *Manduca sexta*: localization and movement during the larval-pupal transformation. *J. Insect Physiol.* **36**, 219–229.

Buckner, J. S., Caldwell, J. M. and Knoper, J. A. (1985). Subcellular localization of uric acid storage in the fat body of *Manduca sexta* during the larval-pupal transformation. *J. Insect Physiol.* **31**, 741–753.

Buckner, J. S., Henderson, T. A., Ehresmann, D. D. and Graf, G. (1990). Structure and composition of urate storage granules from the fat body of *Manduca sexta*. *Insect Biochem.* **20**, 203–214.

Bursell, E. (1965). Nitrogenous waste products of the tsetse fly, *Glossina morsitans*. *J. Insect Physiol.* **11**, 993–1001.

Buss, D. S., McCaffery, A. R. and Callaghan, A. (2002). Evidence for P-glycoprotein modification of insecticide toxicity in mosquitoes of the *Culex pipiens* complex. *Med. Vet. Entomol.* **16**, 218–222.

Byrne, D. N. and Miller, W. B. (1990). Carbohydrate and amino acid composition of phloem sap and honeydew produced by *Bemisia tabaci*. *J. Insect Physiol.* **36**, 433–439.

Cannings, S. C. (1981). The influence of salinity on the cuticular permeability of *Cenocorixa bifida hungerfordi* Lansbury (Hemiptera: Corixidae). *Can. J. Zool.* **59**, 1505–1509.

Chamberlin, M. E. and Phillips, J. E. (1982). Regulation of haemolymph amino acid levels and active secretion of proline by Malpighian tubules of locusts. *Can. J. Zool.* **60**, 2745–2752.

Chintapalli, V. R., Wang, J. and Dow, J. A. (2007). Using FlyAtlas to identify better *Drosophila melanogaster* models of human disease. *Nat. Genet.* **39**, 715–720.

Coast, G. (2007). The endocrine control of salt balance in insects. *Gen. Comp. Endocrinol.* **152**, 332–338.

Coast, G. M., Orchard, I., Phillips, J. E. and Schooley, D. A. (2002). Insect diuretic and antidiuretic hormones. *Adv. Insect Physiol.* **29**, 279–409.

Cochran, D. G. (1985). Comprehensive insect physiology, biochemistry and pharmacology. In: *Nitrogenous Excretion* (eds Kerkut, G. A. and Gilbert, L. I.), pp. 467–506. London: Pergamon.

Collier, K. A. and O'Donnell, M. J. (1997). Analysis of epithelial transport by measurement of K^+, Cl^- and pH gradients in extracellular unstirred layers: ion secretion and reabsorption by Malpighian tubules of an insect, *Rhodnius prolixus*. *J. Exp. Biol.* **200**, 1627–1638.

Cooper, P. D., Scudder, G. G. E. and Quamme, G. A. (1987). Ion and CO_2 regulation in the freshwater water boatman, *Cenocorixa blaisdelli* (Hung.) (Hemiptera, Corixidae). *Physiol. Zool.* **60**, 465–471.

Cooper, P. D., Scudder, G. G. E. and Quamme, G. A. (1989). Segmental differences in secretion by the Malpighian tubules of the fresh water dwelling corixid, *Cenocorixa blaisdelli* (Hung.) (Corixidae, Hemiptera). *J. Insect Physiol.* **35**, 531–536.

Crafts-Brandner, S. J. (2002). Plant nitrogen status rapidly alters amino acid metabolism and excretion in *Bemisia tabaci*. *J. Insect Physiol.* **48**, 33–41.

Craig, A., Hare, L., Charest, P. M. and Tessier, A. (1998). Effect of exposure regime on the internal distribution of cadmium in *Chironomus staegeri* larvae (Insecta, Diptera). *Aquat. Toxicol.* **41**, 265–275.

Craig, A., Hare, L. and Tessier, A. (1999). Experimental evidence for cadmium uptake via calcium channels in the aquatic insect *Chironomus staegeri*. *Aquat. Toxicol.* **44**, 255–262.

Dalton, T. and Windmill, D. M. (1981). The permeability characteristics of the isolated Malphigian tubules of the housefly *Musca domestica*. *Comp. Biochem. Physiol.* **69A**, 211–217.

David, W. A. L. and Gardiner, B. O. C. (1953). Systemic insecticidal action of nicotine and certain other organic bases. *Ann. Appl. Biol.* **40**, 91–105.

Davies, S. A. and Day, J. P. (2006). cGMP signalling in a transporting epithelium. *Biochem. Soc. Trans.* **34**, 512–514.

Davies, S. A., Goodwin, S. F., Kelly, D. C., Wang, Z., Sozen, M. A., Kaiser, K. and Dow, J. A. (1996). Analysis and inactivation of vha55, the gene encoding the vacuolar ATPase B-subunit in *Drosophila melanogaster* reveals a larval lethal phenotype. *J. Biol. Chem.* **271**, 30677–30684.

Day, J. P., Houslay, M. D. and Davies, S. A. (2006). A novel role for a *Drosophila* homologue of cGMP-specific phosphodiesterase in the active transport of cGMP. *Biochem. J.* **393**, 481–488.

Deeley, R. G., Westlake, C. and Cole, S. P. (2006). Transmembrane transport of endo- and xenobiotics by mammalian ATP-binding cassette multidrug resistance proteins. *Physiol. Rev.* **86**, 849–899.

Delaunay, H. (1931). L'excretion azotée des invertébrés. *Biol. Rev.* **6**, 265–301.

Dijkstra, S., Lohrmann, E., Van Kerkhove, E. and Greger, R. (1994). Characteristics of the luminal proton pump in Malpighian tubules of the ant. *Ren. Physiol. Biochem.* **17**, 27–39.

Doichuanngam, K. and Thornhill, R. A. (1992). Penetration, excretion and metabolism of ^{14}C malathion in susceptible and resistant strains of *Plutella xylostella*. *Comp. Biochem. Physiol., C* **101**, 583–588.

Donini, A. and O'Donnell, M. J. (2005). Analysis of Na^+, Cl^-, K^+, H^+ and NH_4^+ concentration gradients adjacent to the surface of anal papillae of the mosquito *Aedes aegypti*: application of self-referencing ion-selective microelectrodes. *J. Exp. Biol.* **208**, 603–610.

Donini, A., Patrick, M. L., Bijelic, G., Christensen, R. J., Ianowski, J. P., Rheault, M. R. and O'Donnell, M. J. (2006). Secretion of water and ions by Malpighian tubules of larval mosquitoes: effects of diuretic factors, 2nd messengers and salinity. *Physiol. Biochem. Zool.* **79**, 645–655.

Douglas, A. E. (1993). The nutritional quality of phloem sap utilized by natural aphid populations. *Ecol. Entomol.* **18**, 31–38.

Douglas, A. E. (1998). Nutritional interactions in insect-microbial symbiosis: aphids and their symbiotic bacteria *Buchnera*. *Ann. Rev. Entomol.* **43**, 17–37.

Dow, J. A. (1999). The multifunctional *Drosophila melanogaster* V-ATPase is encoded by a multigene family. *J. Bioenerg. Biomembr.* **31**, 75–83.

Dow, J. A. and Davies, S. A. (2003). Integrative physiology and functional genomics of epithelial function in a genetic model organism. *Physiol. Rev.* **83**, 687–729.

Dow, J. A. and Davies, S. A. (2006). The Malpighian tubule: rapid insights from post-genomic biology. *J. Insect Physiol.* **52**, 365–378.

Dow, J. A., Davies, S. A., Guo, Y., Graham, S., Finbow, M. E. and Kaiser, K. (1997). Molecular genetic analysis of V-ATPase function in *Drosophila melanogaster*. *J. Exp. Biol.* **200**, 237–245.

Drose, S. and Altendorf, K. (1997). Bafilomycins and concanamycins as inhibitors of V-ATPases and P-ATPases. *J. Exp. Biol.* **200**, 1–8.

Dube, K. A., McDonald, D. G. and O'Donnell, M. J. (2000a). Calcium homeostasis in larval and adult *Drosophila melanogaster*. *Archiv. Insect Biochem. Physiol.* **44**, 27–39.

Dube, K. A,, McDonald, D. G. and O'Donnell, M. J. (2000b). Calcium transport by isolated anterior and posterior Malpighian tubules of *Drosophila melanogaster*: roles of sequestration and secretion. *J. Insect Physiol.* **46**, 1449–1460.

Dubreuil, R. R. (2004). Copper cells and stomach acid secretion in the *Drosophila* midgut. *Int. J. Biochem. Cell Biol.* **36**, 745–752.

Duffey, S. S. and Scudder, G. G. E. (1974). Cardiac glycosides in *Oncopeltus fasciatus* (Dallas) (Hemiptera: Lygaeidae). I. The uptake and distribution of natural cardenolides in the body. *Can. J. Zool.* **52**, 283–290.

Duffey, S. S., Blum, M. S., Isman, M. B. and Scudder, G. G. E. (1978). Cardiac glycosides: a physical system for their sequestration by the milkweed bug. *J. Insect Physiol.* **24**, 639–645.

Echevarria, M., Ramirez-Lorca, R., Hernandez, C. S., Gutierrez, A., Mendez-Ferrer, S., Gonzalez, E., Toledo-Aral, J. J., Ilundain, A. A. and Whittembury, G. (2001). Identification of a new water channel (Rp-MIP) in the Malpighian tubules of the insect *Rhodnius prolixus*. *Pflug. Archiv.* **442**, 27–34.

Ehresmann, D. D., Buckner, J. S. and Graf, G. (1990). Uric acid translocation from the fat body of *Manduca sexta* during the pupal-adult transformation: effects of 20-hydroxyecdysone. *J. Insect Physiol.* **36**, 173–180.

Evans, J. M., Allan, A. K., Davies, S. A. and Dow, J. A. (2005). Sulphonylurea sensitivity and enriched expression implicate inward rectifier K^+ channels in *Drosophila melanogaster* renal function. *J. Exp. Biol.* **208**, 3771–3783.

Fabrick, J. A., Kanost, M. R. and Baker, J. E. (2004). RNAi-induced silencing of embryonic tryptophan oxygenase in the pyralid moth, *Plodia interpunctella*. *J. Insect Sci.* **4**, 15.

Farquharson, P. A. (1974). A study of the Malpighian tubules of the pill millipede, *Glomeris marginata* (Viller3). III. The permeability characteristics of the tubule. *J. Exp. Biol.* **60**, 41–51.

Feyereisen, R. (1999). Insect P450 enzymes. *Annu. Rev. Entomol.* **44**, 507–533.

Frankfater, C., Schuhly, W., Fronczek, F. R. and Slattery, M. (2005). Processing of a sesquiterpene lactone by *Papilio glaucus* caterpillars. *J. Chem. Ecol.* **31**, 2541–2550.

Friedman, T. B. and Johnson, D. H. (1977). Temporal control of urate oxidase activity in *Drosophila*: evidence of an autonomous timer in Malpighian tubules. *Science* **197**, 477–479.

Gäde, G. (1992). The hormonal integration of insect flight metabolism. *Zool. Jahrb.* **96**, 211–225.

Gade, G., Hoffmann, K. H. and Spring, J. H. (1997). Hormonal regulation in insects: facts, gaps, and future directions. *Physiol. Rev.* **77**, 963–1032.

Gaertner, L. S. and Morris, C. E. (1999). Accumulation of daunomycin and fluorescent dyes by drug-transporting Malpighian tubule cells of the tobacco hornworm, *Manduca sexta*. *Tissue Cell* **31**, 185–194.

Gaertner, L. S., Murray, C. L. and Morris, C. E. (1998). Transepithelial transport of nicotine and vinblastine in isolated malpighian tubules of the tobacco hornworm (*Manduca sexta*) suggests a P-glycoprotein-like mechanism. *J. Exp. Biol.* **201**, 2637–2645.

Garayoa, M., Villaro, A. C., Klein, U., Zimmermann, B., Montuenga, L. M. and Sesma, P. (1995). Immunocytochemical localization of a vacuolar-type-ATPase in Malpighian tubules of the ant *Formica polyctena*. *Cell Tissue Res.* **282**, 343–350.

Garcia, E. S., Subrahamanyam, B., Müller, T. and Rembold, H. (1989). Absorption, storage, organ distribution and excretion of dietary (22, 23-$^3H_2$)-dihydroazadirachtin A in the blood-feeding bug, *Rhodnius prolixus*. *J. Insect Physiol.* **35**, 743–749.

Geiser, D. L., Chavez, C. A., Flores-Munguia, R., Winzerling, J. J. and Pham, D. Q. (2003). *Aedes aegypti ferritin*. *Eur. J. Biochem.* **270**, 3667–3674.

Geiser, D. L., Zhang, D. and Winzerling, J. J. (2006). Secreted ferritin: mosquito defense against iron overload? *Insect Biochem. Mol. Biol.* **36**, 177–187.

Giannakou, M. E. and Dow, J. A. T. (2001). Characterization of the *Drosophila melanogaster* alkali-metal/proton exchanger (NHE) gene family. *J. Exp. Biol.* **204**, 3703–3716.

Goldstrohm, D. A., Pennington, J. E. and Wells, M. A. (2003). The role of haemolymph proline as a nitrogen sink during blood meal digestion by the mosquito *Aedes aegypti*. *J. Insect Physiol.* **49**, 115–121.

Graca-Souza, A. V., Maya-Monteiro, C., Paiva-Silva, G. O., Braz, G. R., Paes, M. C., Sorgine, M. H., Oliveira, M. F. and Oliveira, P. L. (2006). Adaptations against heme toxicity in blood-feeding arthropods. *Insect Biochem. Mol. Biol.* **36**, 322–335.

Grodowitz, M. J., Broce, A. B. and Kramer, K. J. (1987). Morphology and biochemical composition of mineralized granules from the Malpighian tubules of *Musca autumnalis* de Geer larvae (Diptera: Muscidae). *Insect Biochem.* **17**, 335–345.

Groenendijk, D., Lucker, S. M., Plans, M., Kraak, M. H. and Admiraal, W. (2002). Dynamics of metal adaptation in riverine chironomids. *Environ. Pollut.* **117**, 101–109.

Gutierrez, A. M., Hernandez, C. S. and Whittembury, G. (2004). A model for fluid secretion in *Rhodnius* upper Malpighian tubules (UMT). *J. Membr. Biol.* **202**, 105–114.

Haley, C. A. and O'Donnell, M. J. (1997). Potassium reabsorption by the lower Malpighian tubule of *Rhodnius prolixus*: inhibition by Ba^{2+} and blockers of H^+/K^+-ATPases. *J. Exp. Biol.* **200**, 139–147.

Haley, C. A., Fletcher, M. and O'Donnell, M. J. (1997). KCl reabsorption by the lower Malpighian tubule of *Rhodnius prolixus*: inhibition by Cl^- channel blockers and acetazolamide. *J. Insect Physiol.* **43**, 657–665.

Hanna, M. A. and Atallah, Y. H. (1971). Penetration and biodegradation of carbaryl in susceptible and resistant strains of the Egyptian cotton leafworm. *J. Econ. Entomol.* **64**, 1391–1395.

Harrison, J. F. and Phillips, J. E. (1992). Recovery from acute haemolymph acidosis in unfed locusts: II. Role of ammonium and titratable acid excretion. *J. Exp. Biol.* **165**, 97–110.

Harvey, W. R. and Wieczorek, H. (1997). Animal plasma membrane energization by chemiosmotic H^+ V-ATPases. *J. Exp. Biol.* **200**, 203–216.

Haunerland, N. H. and Bowers, W. S. (1985). Comparative studies on pharmacokinetics and metabolism of anti-juvenile hormone precocene II. *Archiv. Insect Biochem. Physiol.* **2**, 55–63.

Hayashi, Y. (1960). Xanthine dehydrogenase in the silkworm, *Bombyx mori* L. *Nature* **186**, 1053–1054.

Hazelton, S. R., Felgenhauer, B. E. and Spring, J. H. (2001). Ultrastructural changes in the Malpighian tubules of the house cricket, *Acheta domesticus*, at the onset of diuresis: a time study. *J. Morphol.* **247**, 80–92.

Hegarty, J. L., Zhang, B., Pannabecker, T. L., Petzel, D. H., Baustian, M. D. and Beyenbach, K. W. (1991). Dibutyryl cyclic AMP activates bumetanide-sensitive electrolyte transport in Malpighian tubules. *Am. J. Physiol.* **216**, C521–C529.

Hemingway, J. and Ranson, H. (2000). Insecticide resistance in insect vectors of human disease. *Annu. Rev. Entomol.* **45**, 371–391.

Herbst, D. B. and Bradley, T. J. (1989). A Malpighian tubule lime gland in an insect inhabiting alkaline salt lakes. *J. Exp. Biol.* **145**, 63–78.

Herbst, D. B., Conte, F. P. and Brookes, V. J. (1988). Osmoregulation in an alkaline salt lake insect, *Ephydra* (hydropyrus) *hians* Say (Diptera: Ephydridae) in relation to water chemistry. *J. Insect Physiol.* **34**, 903–909.

Hernandez, C. S., Gutierrez, A. M., Vargas-Janzen, A., Noria, F., Gonzalez, E., Ruiz, V. and Whittembury, G. (2001). Fluid secretion in *Rhodnius* upper malpighian tubules (UMT): water osmotic permeabilities and morphometric studies. *J. Membr. Biol.* **184**, 283–290.

Higgins, C. F. and Gottesman, M. M. (1992). Is the multidrug transporter a flippase? *Trends Biochem. Sci.* **17**, 18–21.

Hopkins, T. L. and Lofgren, P. A. (1968). Adenine metabolism and urate storage in the cockroach *Leucophaea maderae*. *J. Insect Physiol.* **14**, 1803–1814.

Ianowski, J. P. and O'Donnell, M. J. (2001). Transepithelial potential in Malpighian tubules of *Rhodnius prolixus*: lumen-negative voltages and the triphasic response to serotonin. *J. Insect Physiol.* **46**, 107–117.

Ianowski, J. P. and O'Donnell, M. J. (2004). Basolateral ion transport mechanisms during fluid secretion by *Drosophila* Malpighian tubules: Na^+ recycling, Na^+:K^+:$2Cl^-$ cotransport and Cl^- conductance. *J. Exp. Biol.* **207**, 2599–2609.

Ianowski, J. P. and O'Donnell, M. J. (2006). Electrochemical gradients for Na^+, K^+, Cl^- and H^+ across the apical membrane in Malpighian (renal) tubule cells of *Rhodnius prolixus*. *J. Exp. Biol.* **209**, 1964–1975.

Ianowski, J. P., Christensen, R. J. and O'Donnell, M. J. (2002). Intracellular ion activities in Malpighian tubule cells of *Rhodnius prolixus*: evaluation of Na^+:K^+:$2Cl^-$ cotransport across the basolateral membrane. *J. Exp. Biol.* **205**, 1645–1655.

Ianowski, J. P., Christensen, R. J. and O'Donnell, M. J. (2004). Na^+ competes with K^+ in bumetanide-sensitive transport by Malpighian tubules of *Rhodnius prolixus*. *J. Exp. Biol.* **207**, 3707–3716.

Isaacson, L. and Nicolson, S. (1989). A reappraisal of the oil-gap technique for the measurement of transtubular potentials in insect epithelia. *J. Exp. Biol.* **141**, 429–440.

Isman, M. B., Proksch, P. and Witte, L. (1987). Metabolism and excretion of acetylchromenes by the migratory grasshopper. *Archiv. Insect Biochem. Physiol.* **6**, 109–120.

Iyengar, S., Arnason, J. T., Philogene, B. J. R., Werstiuk, N. H. and Morand, P. (1990). Comparative metabolism of the phototoxic allelochemical α-terthienyl in three species of lepidopterans. *Pest. Biochem. Physiol.* **37**, 154–164.

Jarial, M. S., Scudder, G. G. E. and Teraguchi, S. (1969). Observations on the labium of Corixidae (Hemiptera). *Can. J. Zool.* **47**, 713–715.

Johnson, D. L., Majak, W. and Benn, M. H. (2001). Excretion of miserotoxin and detoxification of the aglycone by grasshoppers (Orthoptera: Acrididae). *Phytochemistry* **58**, 739–742.

Kamm, C. D., Tawara, J. N. and Stermitz, F. R. (1998). Spruce budworm larval processing of piperidine alkaloids from spruce needles. *J. Chem. Ecol.* **24**, 1153–1160.

Kang'ethe, W., Aimanova, K. G., Pullikuth, A. K. and Gill, S. S. (2007). NHE8 mediates amiloride-sensitive Na^+/H^+ exchange across mosquito Malpighian tubules and catalyzes Na^+ and K^+ transport in reconstituted proteoliposomes. *Am. J. Physiol.* **292**, F1501–F1512.

Karnaky, K. J., Jr., Petzel, D., Sedmerova, M., Gross, A. and Miller, D. S. (2000). Mrp2-like transport of Texas Red by Malpighian tubules of the common American cockroach, *Periplaneta americana*. *Bull. Mt. Des. Isl. Biol. Lab.* **39**, 52–53.

Karnaky, K. J., Jr., Sedmerova, M., Petzel, D., Bridges, J., Boatwright, S. W. and Miller, D. S. (2001). Mrp2-like transport in the Malpighian tubule of the cricket, *Acheta domesticus*. *Bull. Mt. Des. Isl. Biol. Lab.* **40**, 53–55.

Karnaky, K. J., Jr., Hazen-Martin, D. and Miller, D. S. (2003). The xenobiotic transporter, MRP2, in epithelia from insects, sharks, and the human breast: implications for health and disease. *J. Exp. Zool.* **300**, 91–97.

Kaufmann, N., Mathai, J. C., Hill, W. G., Dow, J. A., Zeidel, M. L. and Brodsky, J. L. (2005). Developmental expression and biophysical characterization of a *Drosophila melanogaster* aquaporin. *Am. J. Physiol.* **289**, C397–C407.

Kiceniuk, J. W. and Phillips, J. E. (1974). Magnesium regulation in mosquito larvae (*Aedes campestris*) living in waters of high $MgSO_4$ content. *J. Exp. Biol.* **61**, 749–760.

Kim, Y. K. and Dantzler, W. H. (1995). Relation of membrane potential to basolateral TEA transport in isolated snake proximal renal tubules. *Am. J. Physiol.* **268**, R1539–R1545.

Klein, U. (1992). The insect V-ATPase, a plasma membrane proton pump energizing secondary active transport: immunological evidence for the occurrence of a V-ATPase in insect ion-transporting epithelia. *J. Exp. Biol.* **172**, 345–354.

Knepper, M. A., Good, D. W. and Burg, M. B. (1984). Mechanism of ammonia secretion by cortical collecting ducts of rabbits. *Am. J. Physiol.* **247**, F729–F738.

Knowles, G. (1975a). The reduced glucose permeability of the isolated Malpighian tubules of the blowfly *Calliphora vomitoria*. *J. Exp. Biol.* **62**, 327–334.

Knowles, G. (1975b). The removal of sulphate by the excretory apparatus of the blowfly *Calliphora vomitoria*. *J. Exp. Biol.* **63**, 237–248.

Komnick, H. and Schmitz, M. (1977). Cutane chloridaufnahme aus hypo-osmotischer Konsentration durch di chlorizellen von *Corixa punctata*. *J. Insect Physiol.* **23**, 165–173.

Komoto, N., Yukuhiro, K. and Tamura, T. (1999). Structure and expression of tandemly duplicated xanthine dehydrogenase genes of the silkworm (*Bombyx mori*). *Insect Mol. Biol.* **8**, 73–83.

Koren, G., Woodland, C. and Ito, S. (1998). Toxic digoxin-drug interactions: the major role of renal P-glycoprotein. *Vet. Hum. Toxicol.* **40**, 45–46.

Krueger, R. A., Broce, A. B. and Hopkins, T. L. (1987). Dissolution of granules in the Malpighian tubules of *Musca autumnalis* de greer, during mineralization of the puparium. *J. Insect Physiol.* **33**, 255–263.

Krueger, R. A., Broce, A. B., Hopkins, T. L. and Kramer, K. J. (1988). Calcium transport from Malpighian tubules to puparial cuticle of *Musca autumnalis*. *J. Comp. Physiol. B* **158**, 413–419.

Kuppers, J., Plagemann, A. and Thurm, U. (1986). Uphill transport of water by electro-osmosis. *J. Memb. Biol.* **91**, 107–119.

Kuterbach, D. A. and Walcott, B. (1986a). Iron-containing cells in the honey-bee (*Apis mellifera*). I. Adult morphology and physiology. *J. Exp. Biol.* **126**, 375–387.

Kuterbach, D. A. and Walcott, B. (1986b). Iron-containing cells in the honey-bee (*Apis mellifera*). II. Accumulation during development. *J. Exp. Biol.* **126**, 389–401.

Kuterbach, D. A., Walcott, B., Reeder, R. J. and Frankel, R. B. (1982). Iron containing cells in the honey bee (*Apis mellifera*). *Science* **218**, 695–697.

Kuzhivelil, B. T. and Mohamed, U. V. K. (1987). The concentration of ammonia in the excreta of sixth instar larvae of *Lamida moncusalis* Walker (Pyralidae: Lepidoptera) during development. *Experientia* **43**, 879–880.

Lane, N. J. and Skaer, H. B. (1980). Intercellular junctions in insects. *Adv. Insect Pkysiol.* **15**, 35–213.

Lazar, K. V. and Mohamed, U. V. K. (1979). The excretion of urea by the larvae of *Spodoptera mauritia* Boisd. (Noctuidae: Lepidoptera) during development. *Experientia* **35**, 1468.

Leader, J. P. and O'Donnell, M. J. (2005). Transepithelial transport of fluorescent P-glycoprotein and MRP2 substrates by insect Malpighian tubules: confocal microscopic analysis of secreted fluid droplets. *J. Exp. Biol.* **208**, 4363–4376.

Lembke, H. F. and Cochran, D. G. (1988). Uric acid in the Malpighian tubules of some blattellid cockroaches. *Comp. Biochem. Physiol.* **91A**, 587–597.

Leslie, E. M., Deeley, R. G. and Cole, S. P. (2005). Multidrug resistance proteins: role of P-glycoprotein, MRP1, MRP2, and BCRP (ABCG2) in tissue defense. *Toxicol. Appl. Pharmacol.* **204**, 216–237.

Levinson, G. and Bradley, T. J. (1984). Removal of insect basal laminae using elastase. *Tissue Cell* **16**, 367–375.

Leyssens, A., Dijkstra, S., Van Kerkove, E. and Steels, P. (1994). Mechanisms of K^+ uptake across the basal membrane of Malpighian tubules of *Formica polyctena*: the effects of ions and inhibitors. *J. Exp. Biol.* **195**, 123–145.

Li, X., Schuler, M. A. and Berembaum, M. R. (2002). Jasmonate and salicylate induce expression of herbivore cytochrome P450 genes. *Nature* **419**, 712–715.

Lindqvist, L., Block, M. and Tjalve, H. (1995). Distribution and excretion of Cd, Hg, Methyl-Hg and Zn in the predatory beetle *Pterostichus niger* (Coleoptera: Carabidae). *Environ. Toxicol. Chem.* **14**, 1195–1201.

Linton, S. and O'Donnell, M. J. (1999). Contribution of K^+:Cl^- cotransport and Na^+/K^+-ATPase to basolateral ion transport in Malpighian tubules of *Drosophila melanogaster*. *J. Exp. Biol.* **202**, 1561–1570.

Linton, S. M. and O'Donnell, M. J. (2000). Novel aspects of the transport of organic anions by the Malpighian tubules of *Drosophila melanogaster*. *J. Exp. Biol.* **203**, 3575–3584.

Lison, L. (1937). Études histophysiologiques sur les tubes de Malpighi des insectes. I. Elimination des colorants acides chez les Orthoptères. *Archs. Biol. Paris* **48**, 321–360.

Locke, M. and Nichol, H. (1992). Iron economy in insects: Transport, metabolism, and storage. *Annu. Rev. Entomol.* **37**, 195–215.

Loveridge, J. P. (1975). Studies on the water balance of adult locusts. III. The water balance of non-flying locusts. *Zool. Afr.* **10**, 1–28.

Machin, J. (1983). Water vapor absorption in insects. *Am. J. Physiol.* **244**, R187–R192.

Maddrell, S. H., Gardiner, B. O., Pilcher, D. E. and Reynolds, S. E. (1974). Active transport by insect Malpighian tubules of acidic dyes and of acylamides. *J. Exp. Biol.* **61**, 357–377.

Maddrell, S. H., Whittembury, G., Mooney, R. L., Harrison, J. B., Overton, J. A. and Rodriguez, B. (1991). The fate of calcium in the diet of *Rhodnius prolixus*: storage in concretion bodies in the Malpighian tubules. *J. Exp. Biol.* **157**, 483–502.

Maddrell, S. H. P. (1969). Secretion by the Malpighian tubules of *Rhodnius*. The movements of ions and water. *J. Exp. Biol.* **51**, 71–97.

Maddrell, S. H. P. (1971). The mechanisms of insect excretory systems. *Adv. Insect Physiol.* **8**, 199–331.

Maddrell, S. H. P. (1977). Transport of ions and water in animal tissues. In: *Insect Malpighian tubules* (eds Gupta, B. L., Moreton, R. B., Oschman, J. L. and Wall, B. J.), pp. 541–569. London: Academic Press.

Maddrell, S. H. P. (1981). The functional design of the insect excretory system. *J. Exp. Biol.* **90**, 1–15.

Maddrell, S. H. P. (1991). The fastest fluid-secreting cell known: the upper Malpighian tubule cell of *Rhodnius*. *BioEssays* **13**, 357–362.

Maddrell, S. H. P. and Gardiner, B. O. C. (1974). The passive permeability of insect Malpighian tubules to organic solutes. *J. Exp. Biol.* **60**, 641–652.

Maddrell, S. H. P. and Gardiner, B. O. C. (1975). Induction of transport of organic anions in Malpighian tubules of *Rhodnius*. *J. Exp. Biol.* **63**, 755–761.

Maddrell, S. H. and Gardiner, B. O. C. (1976). Excretion of alkaloids by Malpighian tubules of insects. *J. Exp. Biol.* **64**, 267–281.

Maddrell, S. H. P. and O'Donnell, M. J. (1992). Insect Malpighian tubules: V-ATPase action in ion and fluid transport. *J. Exp. Biol.* **172**, 417–429.

Maddrell, S. H. P. and O'Donnell, M. J. (1993). Gramicidin switches transport in insect epithelia from potassium to sodium. *J. Exp. Biol.* **177**, 287–292.

Maddrell, S. H. P. and Overton, J. A. (1988). Stimulation of sodium transport and fluid secretion by ouabain in an insect Malpighian tubule. *J. Exp. Biol.* **137**, 265–276.

Maddrell, S. H. P. and Phillips, J. E. (1975a). Active transport of sulphate ions by the Malpighian tubules of the larvae of the mosquito *Aedes campestris*. *J. Exp. Biol.* **62**, 367–378.

Maddrell, S. H. P. and Phillips, J. E. (1975b). Secretion of hypoosmotic fluid by the lower Malpighian tubules of *Rhodnius prolixus*. *J. Exp. Biol.* **62**, 671–683.

Maddrell, S. H. P. and Phillips, J. E. (1978). Induction of sulphate sransport and hormonal control of fluid secretion by Malpighian tubules of larvae of the mosquito, *Aedes taeniorhynchus*. *J. Exp. Biol.* **72**, 181–202.

Maddrell, S. H. P., O'Donnell, M. J. and Caffrey, R. (1993). The regulation of haemolymph potassium activity during initiation and maintenance of diuresis in fed *Rhodnius prolixus*. *J. Exp. Biol.* **177**, 273–285.

Majak, W., Johnson, D. L. and Benn, M. H. (1998). Detoxification of 3-nitropropionic acid and karakin by melanopline grasshoppers. *Phytochem.* **49**, 419–422.

Maroni, G. and Watson, D. (1985). Uptake and binding of cadmium, copper and zinc in *Drosophila melanogaster* larvae. *Insect Biochem.* **15**, 55–63.

Marshall, A. T. and Wood, R. W. (1990). Ionic and osmotic regulation by larvae of the sheep blowfly *Lucilia cuprina*. *J. Insect Physiol.* **36**, 635–639.

Marshall, A. T. and Wright, A. (1974). Ultrastructure changes associated with osmoregulation in the hindgut cells of a saltwater insect, *Ephydrell* sp. (ephydridae, diptera). *Tissue Cell* **6**, 301–308.

Marshall, A. T., Cooper, P., Rippon, G. D. and Patak, A. E. (1993). Ion and fluid secretion by different segments of the Malpighian tubules of the black field cricket *Teleogryllus oceanicus*. *J. Exp. Biol.* **177**, 1–22.

Martinez-Barnetche, J., Garcia Solache, M., Neri Lecona, A., Tello Lopez, A. T., del Carmen Rodriguez, M., Gamba, G., Vazquez, N., Rodriguez, M. H. and Lanz-Mendoza, H. (2007). Cloning and functional characterization of the *Anopheles albimanus* DMT1/NRAMP homolog: implications in iron metabolism in mosquitoes. *Insect Biochem. Mol. Biol.* **37**, 532–539.

Martini, S. V., Goldenberg, R. C., Fortes, F. S., Campos-de-Carvalho, A. C., Falkenstein, D. and Morales, M. M. (2004). *Rhodnius prolixus* Malpighian tubule's aquaporin expression is modulated by 5-hydroxytryptamine. *Archiv. Insect Biochem. Physiol.* **57**, 133–141.

Masia, R., Aneshansley, D., Nagel, W., Nachman, R. J. and Beyenbach, K. W. (2000). Voltage clamping single cells in intact Malpighian tubules of mosquitoes. *Am. J. Physiol. Renal Physiol.* **279**, F744–F754.

Meredith, J. and Phillips, J. E. (1973). Rectal ultrastructure in salt- and freshwater mosquito larvae in relation to physiological state. *Cell Tissue Res.* **138**, 1–22.

Meredith, J., Moore, L. and Scudder, G. G. E. (1984). Excretion of ouabain by Malpighian tubules of *Oncopeltus fasciatus*. *Am. J. Physiol.* **246**, R705–R715.

Meredith, J., Ring, M., Macins, A., Marschall, J., Cheng, N. N., Theilmann, D., Brock, H. W. and Phillips, J. E. (1996). Locust ion transport peptide (ITP): primary structure, cDNA and expression in a baculovirus system. *J. Exp. Biol.* **199**, 1053–1061.

Merzendorfer, H., Graf, R., Huss, M., Harvey, W. R. and Wieczorek, H. (1997). Regulation of proton-translocating V-ATPases. *J. Exp. Biol.* **200**, 225–235.

Meulemans, W. and De Loof, A. (1992). Transport of the cationic fluorochrome rhodamine 123 in an insect's Malpighian tubule: indications of a reabsorptive function of the secondary cell type. *J. Cell Sci.* **101**, 349–361.

Miles, P. W. (1966). A modification of Wigglesworth's model for the excretion of uric acid in insects, in the light of modern hypotheses of ion transport. *J. Theor. Biol.* **12**, 130–132.

Montoreano, R., Triana, F., Abate, T. and Rangel-Aldao, R. (1990). Cyclic AMP in the Malpighian tubule fluid and in the urine of *Rhodnius prolixus*. *Gen. Comp. Endocrinol.* **77**, 136–142.

Mordue, A. J. and Blackwell, A. (1993). Azadirachtin: an update. *J. Insect Physiol.* **39**, 903–924.

Mullins, D. E. (1974). Nitrogen metabolism in the American cockroach: an examination of whole body ammonium and other cations excreted in relation to water requirements. *J. Exp. Biol.* **61**, 541–556.

Mullins, D. E. and Cochran, D. G. (1972). Nitrogen excretion in cockroaches: uric acid is not a major product. *Science* **177**, 699–701.

Mullins, D. E. and Cochran, D. G. (1973a). Nitrogenous excretory materials from the American cockroach. *J. Insect Physiol.* **19**, 1007–1018.

Mullins, D. E. and Cochran, D. G. (1973b). Tryptophan metabolite excretion by the American cockroach. *Comp. Biochem. Physiol., B* **44**, 549–555.

Mullins, D. E. and Cochran, D. G. (1975). Nitrogen metabolism in the American cockroach. II. An examination of negative nitrogen balance with respect to mobilization of uric acid and stores. *Comp. Biochem. Physiol.* **50A**, 501–510.

Murray, C. L., Quaglia, M., Arnason, J. T. and Morris, C. E. (1994). A putative nicotine pump at the metabolic blood-brain barrier of the tobacco hornworm. *J. Neurobiol.* **25**, 23–34.

Nachman, R. J., Strey, A., Isaac, E., Pryor, N., Lopez, J. D., Deng, J. G. and Coast, G. M. (2002). Enhanced in vivo activity of peptidase-resistant analogs of the insect kinin neuropeptide family. *Peptides* **23**, 735–745.

Neufeld, D. S. G., Kauffman, R. and Kurtz, Z. (2005). Specificity of the fluorescein transport process in Malpighian tubules of the cricket *Acheta domesticus*. *J. Exp. Biol.* **208**, 2227–2236.

Nicolson, S. and Isaacson, L. (1990). Patch clamp of the basal membrane of beetle Malpighian tubules: direct demonstration of potassium channels. *J. Insect Physiol.* **36**, 877–884.

Nicolson, S., Isaacson, L. and Gerneke, D. (1991). A new method of preparing the basal membrane of renal tubules for patch clamp, using beetle malpihian tubules. *Pflug. Archiv.* **417**, 654–656.

Nijhout, H. F. (1975). Excretory role of the midgut in larvae of the tobacco hornworm, *Manduca sexta* (L.). *J. Exp. Biol.* **62**, 221–230.

Nitao, J. K. (1990). Metabolism and excretion of the furanocoumarin xanthotoxin by parsnip webworm, *Depressaria pastinacella*. *J. Chem. Ecol.* **16**, 417–428.

Noble-Nesbitt, J. (1970). Water balance in the Firebrat, *Thermobia domestica* (Packard). The site of uptake of water from the atmosphere. *J. Exp. Biol.* **52**, 193–200.

O'Connor, K. R. and Beyenbach, K. W. (2001). Chloride channels in apical membrane patches of stellate cells of Malpighian tubules of *Aedes aegypti*. *J. Exp. Biol.* **204**, 367–378.

O'Donnell, M. J. and Leader, J. P. (2006). Changes in fluid secretion rate alter net transepithelial transport of MRP2 and P-glycoprotein substrates in Malpighian tubules of *Drosophila melanogaster*. *Archiv. Insect Biochem. Physiol.* **63**, 123–134.

O'Donnell, M. J. and Machin, J. (1988). Water vapor absorption by terrestrial organisms. *Adv. Comp. Environ. Physiol.* **2**, 47–90.

O'Donnell, M. J. and Machin, J. (1991). Ion activities and electrochemical gradients in the mealworm rectal complex. *J. Exp. Biol.* **155**, 375–402.

O'Donnell, M. J. and Maddrell, S. H. P. (1984). Secretion by the Malpighian tubules of *Rhodnius prolixus* Stal: electrical events. *J. Exp. Biol.* **110**, 275–290.

O'Donnell, M. J. and Maddrell, S. H. P. (1995). Fluid reabsorption and ion transport by the lower Malpighian tubules of adult female *Drosophila*. *J. Exp. Biol.* **198**, 1643–1647.

O'Donnell, M. J. and Rheault, M. R. (2005). Ion-selective microelectrode analysis of salicylate transport by the Malpighian tubules and gut of *Drosophila melanogaster*. *J Exp. Biol.* **208**, 93–104.

O'Donnell, M. J., Aldis, G. K. and Maddrell, S. H. P. (1982). Measurements of osmotic permeability in the Malpighian tubules of an insect, *Rhodnius prolixus* Stal. *Proc. Roy. Soc. Lond. B* **216**, 267–277.

O'Donnell, M. J., Maddrell, S. H. P. and Gardiner, B. O. C. (1983). Transport of uric acid by the Malpighian tubules of *Rhodnius prolixus* and other insects. *J. Exp. Biol.* **103**, 169–184.

O'Donnell, M. J., Maddrell, S. H. P. and Gardiner, B. O. C. (1984). Passage of solutes through the walls of the Malpighian tubules of *Rhodnius* by paracellular and transcellular routes. *Am. J. Physiol.* **246**, R759–R769.

O'Donnell, M. J., Dow, J. A. T., Huesmann, G. R., Tublitz, N. J. and Maddrell, S. H. P. (1996). Separate control of anion and cation transport in Malpighian tubules of *Drosophila melanogaster*. *J. Exp. Biol.* **199**, 1163–1175.

O'Donnell, M. J., Rheault, M. R., Davies, S. A., Rosay, P., Harvey, B. J., Maddrell, S. H. P., Kaiser, K. and Dow, J. A. T. (1998). Hormonally-controlled chloride movement across *Drosophila* tubules is via ion channels in stellate cells. *Am. J. Physiol.* **43**, R1039–R1049.

O'Donnell, M. J., Ianowski, J. P., Linton, S. M. and Rheault, M. R. (2003). Inorganic and organic anion transport by insect renal epithelia. *Biochim. Biophys. Acta* **1618**, 194–206.

Orchard, I. (2006). Serotonin: a coordinator of feeding-related physiological events in the blood-gorging bug, *Rhodnius prolixus*. *Comp. Biochem. Physiol. A* **144**, 316–324.

Paiva-Silva, G. O., Cruz-Oliveira, C., Nakayasu, E. S., Maya-Monteiro, C. M., Dunkov, B. C., Masuda, H., Almeida, I. C. and Oliveira, P. L. (2006). A haeme-degradation pathway in a blood-sucking insect. *Proc. Natl. Acad. Sci. USA* **103**, 8030–8035.

Pasteur, N., Nance, E. and Bons, N. (2001). Tissue localization of overproduced esterases in the mosquito *Culex pipiens* (Diptera: Culicidae). *J. Med. Entomol.* **38**, 791–801.

Patrick, M. L. and Bradley, T. J. (2000a). The physiology of salinity tolerance in larvae of two species of *Culex* mosquitoes: the role of compatible solutes. *J. Exp. Biol.* **203**, 821–830.

Patrick, M. L. and Bradley, T. J. (2000b). Regulation of compatible solute accumulation in larvae of the mosquito *Culex tarsalis*: osmolarity versus salinity. *J. Exp. Biol.* **203**, 831–839.

Patrick, M. L., Aimanova, K., Sanders, H. R. and Gill, S. S. (2006). P-type Na^+/K^+-ATPase and V-type H^+-ATPase expression patterns in the osmoregulatory organs of larval and adult mosquito *Aedes aegypti*. *J. Exp. Biol.* **209**, 4638–4651.

Peach, J. L. and Phillips, J. E. (1991). Metabolic support of chloride-dependent short-circuit current across the locust (*Schistocerca gregaria*) ileum. *J. Insect Physiol.* **37**, 255–260.

Petzel, D. H. (2000). Na^+/H^+ exchange in mosquito Malpighian tubules. *Am. J. Physiol.* **279**, R1996–R2003.

Phillips, J. E. (1964a). Rectal absorption in the desert locust, *Schistocerca gregaria* Forskål. I. Water. *J. Exp. Biol.* **41**, 15–38.

Phillips, J. E. (1964b). Rectal absorption in the desert locust, *Schistocerca gregaria* Forskål. II. Sodium, potassium and chloride. *J. Exp.Biol.* **41**, 39–67.

Phillips, J. E. (1977). Excretion in insects: function of gut and rectum in concentrating and diluting the urine. *Fed. Proc.* **36**, 2480–2486.

Phillips, J. E. (1981). Comparative physiology of insect renal function. *Am. J. Physiol.* **241**, R241–R257.

Phillips, J. E. and Dockrill, A. A. (1968). Molecular sieving of hydrophilic molecules by the rectal intima of the desert locust (*Schistocerca gregaria*). *J. Exp. Biol.* **48**, 521–532.

Phillips, J. E. and Maddrell, S. H. P. (1974). Active transport of magnesium by the Malpighian tubules of the larvae of the mosquito, *Aedes campestris*. *J. Exp. Biol.* **61**, 761–771.

Phillips, J. E., Hanrahan, J., Chamberlin, M. and Thomson, B. (1986). Mechanisms and control of reabsorption in insect hindgut. *Adv. Insect Physiol.* **19**, 329–422.

Phillips, J. E., Wiens, C., Audsley, N., Jeffs, L., Bilgen, T. and Meredith, J. (1996). Nature and control of chloride transport in insect absorptive epithelia. *J. Exp. Zool.* **275**, 292–299.

Plapp, F. W., Bigley, W. S., Darrow, D. I. and Eddy, G. W. (1961). Studies on parathion metabolism in normal and parathion resistant houseflies. *J. Econ. Entomol.* **54**, 389–392.

Podsiadlowski, L., Matha, V. and Vilcinskas, A. (1998). Detection of a P-glycoprotein related pump in *Chironomus* larvae and its inhibition by verapamil and cyclosporin A. *Comp. Biochem. Physiol., B* **121**, 443–450.

Polles, S. G. and Vinson, S. B. (1972). Penetration, distribution and metabolism of 14C endrin in resistant and susceptible tobacco budworm larvae. *J. Agr. Food Chem.* **20**, 38–41.

Pond, B. B., Galeffi, F., Ahrens, R. and Schwartz-Bloom, R. D. (2004). Chloride transport inhibitors influence recovery from oxygen-glucose deprivation-induced cellular injury in adult hippocampus. *Neuropharmacology* **47**, 253–262.

Postma, J. F., van Nugteren, P. and Buckert-de Jong, M. B. (1996). Increased cadmium excretion in metal-adapted populations of the midge *Chironomus riparius* (Diptera). *Environ. Toxicol. Chem.* **15**, 332–339.

Pritchard, J. B. and Miller, D. S. (1991). Comparative insights into the mechanisms of renal organic anion and cation secretion. *Am. J. Physiol.* **261**, R1329–R1340.

Pritchard, J. B. and Miller, D. S. (1993). Mechanisms mediating renal secretion of organic anions and cations. *Physiol. Rev.* **73**, 765–796.

Prusch, R. D. (1972). Secretion of NH_4Cl by the hindgut of *Sarcophaga bullata* larvae. *Comp. Biochem. Physiol.* **41A**, 215–223.

Prusch, R. D. (1976). Unidirectional ion movements in the hindgut of larval *Sarcophaga bullata*. *J. Exp. Biol.* **64**, 89–100.

Pullikuth, A. K., Filippov, V. and Gill, S. S. (2003). Phylogeny and cloning of ion transporters in mosquitoes. *J. Exp. Biol.* **206**, 3857–3868.

Pullikuth, A. K., Aimanova, K., Kang'ethe, W., Sanders, H. R. and Gill, S. S. (2006). Molecular characterization of sodium/proton exchanger 3 (NHE3) from the yellow fever vector, *Aedes aegypti*. *J. Exp. Biol.* **209**, 3529–3544.

Rabitsch, W. B. (1997). Tissue-specific accumulation patterns of Pb, Cd, Cu, Zn, Fe, and Mn in workers of three ant species (Formicidae, Hymenoptera) from a metal-polluted site. *Archiv. Environ. Contam. Toxicol.* **32**, 172–177.

Rafaeli-Bernstein, A. and Mordue, W. (1978). The transport of the cardiac glycoside ouabain by the Malpighian tubules of *Zonocerus variegatus*. *Physiol. Entomol.* **3**, 59–63.

Ramsay, J. A. (1955a). The excretory system of the stick insect, *Dixippus morosus* (Orthoptera, Phasmidae). *J. Exp. Biol.* **32**, 183–199.

Ramsay, J. A. (1955b). The excretion of sodium, potassium and water by the Malpighian tubules of the stick insect, *Dixippus morosus* (Orthoptera, Phasmidae). *J. Exp. Biol.* **32**, 200–216.

Ramsay, J. A. (1958). Excretion by the Malpighian tubules of the stick insect, *Dixippus morosus* (Orthoptera, Phasmidae): amino acids, sugars and urea. *J. Exp. Biol.* **35**, 871–891.

Ramsay, J. A. (1971). Insect rectum. *Phil. Trans. R. Soc. Lond. B* **262**, 251–260.

Ranson, H. and Hemingway, J. (2005). Mosquito glutathione transferases. *Methods Enzymol.* **401**, 226–241.

Reagan, J. D. (1995). Molecular cloning of a putative Na^+–K^+–$2Cl^-$ cotransporter from the Malpighian tubules of the tobacco hornworm, *Manduca sexta*. *Insect Biochem. Mol. Biol.* **25**, 875–880.

Reaume, A. G., Knecht, D. A. and Chovnick, A. (1991). The rosy locus in *Drosophila melanogaster*: xanthine dehydrogenase and eye pigments. *Genetics* **129**, 1099–1109.

Rheault, M. R. and O'Donnell, M. J. (2004). Organic cation transport by Malpighian tubules of *Drosophila melanogaster:* application of two novel electrophysiological methods. *J. Exp. Biol.* **207**, 2173–2184.

Rheault, M. R., Debicki, D. M. and O'Donnell, M. J. (2005). Characterization of tetraethylammonium uptake across the basolateral membrane of the *Drosophila* Malpighian (renal) tubule. *Am. J. Physiol.* **289**, R495–R504.

Rheault, M. R., Plaumann, J. S. and O'Donnell, M. J. (2006). TEA and nicotine transport by the Malpighian tubules of insects. *J. Insect Physiol.* **52**, 487–498.

Rheault, M. R., Okech, B. A., Keen, S. B., Miller, M. M., Meleshkevitch, E. A., Linser, P. J., Boudko, D. Y. and Harvey, W. R. (2007). Molecular cloning, phylogeny and localization of AgNHA1: the first Na^+/H^+ antiporter (NHA) from a metazoan, *Anopheles gambiae*. *J Exp. Biol.* **210**, 3848–3861.

Riegel, J. A., Maddrell, S. H., Farndale, R. W. and Caldwell, F. M. (1998). Stimulation of fluid secretion of malpighian tubules of *Drosophila melanogaster* Meig. by cyclic nucleotides of inosine, cytidine, thymidine and uridine. *J. Exp. Biol.* **201**, 3411–3418.

Riegel, J. A., Farndale, R. W. and Maddrell, S. H. (1999). Fluid secretion by isolated Malpighian tubules of *Drosophila melanogaster* Meig.: effects of organic anions, quinacrine and a diuretic factor found in the secreted fluid. *J. Exp. Biol.* **202**, 2339–2348.

Riskallah, M. R. and Esaac, E. G. (1982). Metabolism and excretion of leptophos and its phenol in S and R larvae of the Egyptian cotton leafworm. *Archiv. Environ. Contam. Toxicol.* **11**, 253–257.

Ruiz-Sanchez, E. and O'Donnell, M. J. (2006). Characterization of salicylate uptake across the basolateral membrane of the Malpighian tubules of *Drosophila melanogaster. J. Insect Physiol.* **52**, 920–928.

Ruiz-Sanchez, E. and O'Donnell, M. J. (2007a). Characterization of transepithelial transport of salicylate by the Malpighian tubules of *Drosophila melanogaster* and the effects of changes in fluid secretion rate. *Physiol. Entomol.* **32**, 157–166.

Ruiz-Sanchez, E. and O'Donnell, M. J. (2007b). Effects of chronic exposure to dietary salicylate on elimination and renal excretion of salicylate by *Drosophila melanogaster* larvae. *J. Exp. Biol.* **210**, 2464–2471.

Ruiz-Sanchez, E., Livingston, A., Van Walderveen, M. C. and O'Donnell, M. J. (2007). Transepithelial transport of salicylate by the Malpighian tubules of insects from different orders. *J. Insect Physiol.* **53**, 1034–1045.

Russell, J. M. (2000). Sodium-potassium-chloride cotransporter. *Phys. Rev.* **80**, 211–276.

Sasaki, T. and Ishikawa, H. (1995). Production of essential amino acids from glutamate by Mycetocyte symbionts of the pea aphid, *Acyrthosiphon pisum. J. Insect Physiol.* **41**, 41–46.

Sawyer, D. B. and Beyenbach, K. W. (1985). Dibutyryl-cAMP increases basolateral sodium conductance of mosquito Malpighian tubules. *Am. J. Physiol.* **248**, R339–R345.

Scaraffia, P. Y., Isoe, J., Murillo, A. and Wells, M. A. (2005). Ammonia metabolism in *Aedes aegypti. Insect Biochem. Mol. Biol.* **35**, 491–503.

Schofield, R. M., Postlethwait, J. H. and Lefevre, H. W. (1997). MeV-ion microprobe analyses of whole *Drosophila* suggest that zinc and copper accumulation is regulated storage not deposit excretion. *J. Exp. Biol.* **200**, 3235–3243.

Schwantes, U. (1990). Uric acid during pupal and adult development of *Musca domestica* L. (Diptera). *Zool. Jb. Physiol.* **94**, 1–18.

Sciortino, C. M., Shrode, L. D., Fletcher, B. R., Harte, P. J. and Romero, M. F. (2001). Localization of endogenous and recombinant Na^+-driven anion exchanger protein NDAE1 from *Drosophila melanogaster. Am. J. Physiol.* **281**, C449–C463.

Scudder, G. G. E. (1969). The fauna of saline lakes of the Fraser plateau in British Columbia. *Vekr. Intemat. Verein. Lirnnol.* **17**, 430–439.

Seidman, L. A., Bergtom, G. and Remsen, C. C. (1986a). Structure of the larval midgut of the fly *Chironomus thummi* and its relationship to sites of cadmium sequestration. *Tissue Cell* **18**, 407–418.

Seidman, L. A., Bergtrom, G., Gingrich, D. J. and Remsen, C. C. (1986b). Accumulation of cadmium by the fourth instar larva of the fly *Chironomus thummi. Tissue Cell* **18**, 395–405.

Self, L. S., Guthrie, F. E. and Hodgson, E. (1964). Adaptation of tobacco hornworms to the ingestion of nicotine. *J. Insect Physiol.* **10**, 907–914.

Skaer, H. B., Maddrell, S. H. and Harrison, J. B. (1987). The permeability properties of septate junctions in Malpighian tubules of *Rhodnius. J. Cell Sci.* **88**, 251–265.

Smith, P. J., Sanger, R. H. and Jaffe, L. F. (1994). The vibrating Ca^{2+} electrode: a new technique for detecting plasma membrane regions of Ca^{2+} influx and efflux. *Methods Cell Biol.* **40**, 115–134.

Smitherman, P. K., Townsend, A. J., Kute, T. E. and Morrow, C. S. (2004). Role of multidrug resistance protein 2 (MRP2, ABCC2) in alkylating agent detoxification: MRP2 potentiates glutathione S-transferase A1-1-mediated resistance to chlorambucil cytotoxicity. *J. Pharmacol. Exp. Ther.* **308**, 260–267.

Snyder, M. J., Walding, J. K. and Feyereisen, R. (1994). Metabolic fate of the allelochemical nicotine in the tobacco hornworm *Manduca sexta*. *Insect Biochem. Mol. Biol.* **24**, 837–846.

Sohal, R. S., Peters, P. D. and Hall, T. A. (1977). Origin, structure, composition and age dependence of mineralised dense bodies (concretions) in the midgut epithelium of the adult housefly *Musca domestica*. *Tissue Cell* **9**, 87–102.

Souza, A. V., Petretski, J. H., Demasi, M., Bechara, E. J. and Oliveira, P. L. (1997). Urate protects a blood-sucking insect against hemin-induced oxidative stress. *Free Radic. Biol. Med.* **22**, 209–214.

Spring, J. H. and Felgenhauer, B. E. (1996). Excretion in the house cricket, *Acheta domesticus*: effects of diuretics on the structure of the mid-tubule. *J. Morphol.* **230**, 43–53.

Spring, J. H. and Hazelton, S. R. (1987). Excretion in the house cricket (*Acheta domesticus*): stimulation of diuresis by tissue homogenates. *J. Exp. Biol.* **129**, 63–81.

Spring, J. H., Robichaux, S. R., Kaufmann, N. and Brodsky, J. L. (2007). Localization of a Drosophila DRIP-like aquaporin in the Malpighian tubules of the house cricket, *Acheta domesticus*. *Comp. Biochem. Physiol.* **148**, 92–100.

Staddon, B. W. (1955). The excretion and storage of ammonia by aquatic larvae of *Sialis lutaria* (Neuroptera). *J. Exp. Biol.* **32**, 84–94.

Staddon, B. W. (1959). Nitrogen excretion in nymphs of *Aeshna cyanea* (Müll.) (Odonata, Anisoptera). *J. Exp. Biol.* **36**, 566–574.

Stagg, A. P., Harrison, J. F. and Phillips, J. E. (1991). Acid-base variables in Malpighian tubule secretion and response to acidosis. *J. Exp. Biol.* **159**, 433–447.

Stobbart, R. H. (1967). The effect of some anions and cations upon the fluxes and net uptake of chloride in the larva of *Aedes aegypti* (L.), and the nature of the uptake mechanisms for sodium and chloride. *J. Exp. Biol.* **47**, 35–57.

Storelli, C., Corcelli, A., Cassano, G., Hildmann, B., Murer, H. and Lippe, C. (1980). Polar distribution of sodium-dependent and sodium-independent transport systems for L-lactate in the plasma membrane of rat enterocytes. *Pflug. Archiv.* **388**, 11–16.

Strange, K. and Phillips, J. E. (1984). Mechanisms of CO_2 transport in rectal salt gland of *Aedes*. I. Ionic requirements of CO_2 secretion. *Am. J. Physiol.* **246**, R727–R734.

Strange, K. and Phillips, J. E. (1985). Cellular mechanism of HCO_3^- and Cl^- transport in insect salt gland. *J. Membr. Biol.* **83**, 25–37.

Strange, K., Phillips, J. E. and Quamme, G. A. (1982). Active HCO_3^- secretion in the rectal salt gland of a mosquito larva inhabiting $NaHCO_3$-CO_3 lakes. *J. Exp. Biol.* **101**, 171–186.

Strange, K., Phillips, J. E. and Quamme, G. A. (1984). Mechanisms of CO_2 transport in rectal salt gland of *Aedes*. II. Site of Cl^--HCO_3^- exchange. *Am. J. Physiol.* **246**, R735–R740.

Stratakis, E. (1979). Ommochrome synthesis and kynurenic acid excretion in relation to metamorphosis and allatectomy in the stick insect, *Carausius morosus*. *J. Insect Physiol.* **25**, 925–929.

Suzuki, K. T., Aoki, Y., Nishikawa, M., Masui, H. and Matsubara, F. (1984). Effect of cadmium-feeding on tissue concentrations of elements in germ-free silkworm (*Bombyx mori*) larvae and distribution of cadmium in the alimentary canal. *Comp. Biochem. Physiol., C* **79**, 249–253.

Szibbo, C. M. and Scudder, G. G. E. (1979). Secretory activity of the segmented Malpighian tubules of *Cenocorixa bifida* (Hung.). *J. Insect Physiol.* **25**, 931–937.

Tapadia, M. G. and Lakhotia, S. C. (2005). Expression of MDR49 and MDR65 multidrug resistance genes in larval tissues of *Drosophila melanogaster* under normal and stress conditions. *Cell Stress Chap.* **10**, 7–11.

Taylor, C. W. (1985). Calcium regulation in blowflies: absence of a role for midgut. *Am. J. Physiol.* **249**, R209–R213.

Tearle, R. (1991). Tissue specific effects of ommochrome pathway mutations in *Drosophila melanogaster*. *Genet. Res.* **57**, 257–266.

Terhzaz, S., O'Connell, F. C., Pollock, V. P., Kean, L., Davies, S. A., Veenstra, J. A. and Dow, J. A. (1999). Isolation and characterization of a leucokinin-like peptide of *Drosophila melanogaster*. *J. Exp. Biol.* **202**, 3667–3676.

Thomson, R. B., Thomson, J. M. and Phillips, J. E. (1988). transport in acid-secreting insect epthelium. *Am. J. Physiol.* **254**, R348–R356.

Torrie, L. S., Radford, J. C., Southall, T. D., Kean, L., Dinsmore, A. J., Davies, S. A. and Dow, J. A. (2004). Resolution of the insect ouabain paradox. *Proc. Natl. Acad. Sci. USA* **101**, 13689–13693.

Treherne, J. E., Harrison, J. B., Treherne, J. M. and Lane, N. J. (1984). Glial repair in an insect central nervous system: effects of surgical lesioning. *J. Neurosci.* **4**, 2689–2697.

Tucker, L. E. (1977a). The influence of diet, age and state of hydration on Na^+, K^+ and urate balance in the fat body of the cockroach *Periplaneta americana*. *J. Exp. Biol.* **71**, 67–79.

Tucker, L. E. (1977b). Regulation of ions in the haemolymph of the cockroach *Periplaneta americana* during dehydration and rehydration. *J. Exp. Biol.* **71**, 95–110.

Van Kerkhove, E., Weltens, R., Roinel, N. and De Decker, N. (1989). Haemolymph composition in *Formica* (Hymenoptera) and urine formation by the short isolated Malpighian tubules: electrochemical gradients for ion transport. *J. Insect Physiol.* **35**, 991–1003.

Wall, B. J. (1971). Local osmotic gradients in the rectal pads of an insect. *Fed. Proc.* **30**, 42–48.

Wang, J., Kean, L., Yang, J., Allan, A. K., Davies, S. A., Herzyk, P. and Dow, J. A. (2004). Function-informed transcriptome analysis of *Drosophila* renal tubule. *Genome Biol.* **5**, R69.

Wang, Y. J., Zhao, Y., Meredith, J., Phillips, J. E., Theilmann, D. A. and Brock, H. W. (2000). Mutational analysis of the C-terminus in ion transport peptide (ITP) expressed in *Drosophila* Kc1 cells. *Archiv. Insect Biochem. Physiol.* **45**, 129–138.

Weihrauch, D. (2006). Active ammonia absorption in the midgut of the tobacco hornworm *Manduca sexta* L.: transport studies and mRNA expression analysis of a Rhesus-like ammonia transporter. *Insect Biochem. Mol. Biol.* **36**, 808–821.

Weis-Fogh, T. (1967). Insects and physiology. In: *Metabolism and Weight Economy in Migrating Animals, Particularly Birds and Insects* (eds Beament, J. W. L. and Treherne, J. E.), pp. 143–159. Edinburgh and London: Oliver and Boyd.

Whittembury, G., Biondi, A. C., Paz-Aliaga, A., Linares, H., Parthe, V. and Linares, L. (1986). Transcellular and paracellular plow of water during secretion in the upper segment of the Malpighian tubule of *Rhodnius prolixus*: solvent drag of molecules of graded size. *J. Exp. Biol.* **123**, 71–92.

Wieczorek, H. (1992). The insect V-ATPase, a plasma membrane proton pump energizing secondary active transport: molecular analysis of electrogenic potassium transport in the tobacco hornworm midgut. *J. Exp. Biol.* **172**, 335–343.

Wieczorek, H., Weerth, S., Schindlbeck, M. and Klein, U. (1989). A vacuolar-type proton pump in a vesicle fraction enriched with potassium transporting plasma membranes from tobacco hornworm midgut. *J. Biol. Chem.* **264**, 11143–11148.

Wieczorek, H., Putzenlechner, M., Zeiske, W. and Klein, U. (1991). A vacuolar-type proton pump energizes K^+/H^+ antiport in an animal plasma membrane. *J. Biol. Chem.* **266**, 15340–15347.

Wieczorek, H., Grber, G., Harvey, W. R., Huss, M., Merzendorfer, H. and Zeiske, W. (2000). Structure and regulation of insect plasma membrane H^+ V-ATPase. *J. Exp. Biol.* **203**, 127–135.

Wiehart, U. I., Klein, G., Steels, P., Nicolson, S. W. and Van Kerkhove, E. (2003). K^+ transport in Malpighian tubules of *Tenebrio molitor* L: is a K_{ATP} channel involved? *J. Exp. Biol.* **206**, 959–965.

Wigglesworth, V. B. (1931). The physiology of excretion in a blood-sucking insect, *Rhodnius prolixus* (Hemiptera, Reduviidae). III. The mechanism of uric acid excretion. *J. Exp. Biol.* **8**, 448–451.

Wigglesworth, V. B. (1972). *The Principles of Insect Physiology*, 7th edition. London: Methuen.

Wigglesworth, V. B. (1987a). Histochemical studies of uric acid in some insects. 1. Storage in the fat body of *Periplaneta maericana* and the action of the symbiotic bacteria. *Tissue Cell* **19**, 83–91.

Wigglesworth, V. B. (1987b). Histochemical studies of uric acid in some insects. 3. Excretion of uric acid by the Malpighian tubules in *Calliphora vicina* and *Rhodnius prolixus*. *Tissue Cell* **19**, 275–286.

Wilkinson, T. L. and Douglas, A. E. (1996). The impact of aposymbiosis on amino acid metabolism of pea aphids. *Entomol. Exp. Appl.* **80**, 279–282.

Williams, L. A. D. and Mansingh, A. (1996). The insecticidal and acaricidal actions of compounds from *Azadirachta indica* (A. Juss) and their use in tropical pest management. *Int. Pest Manage. Rev.* **1**, 133–145.

Wink, M. and Theile, V. (2002). Alkaloid tolerance in *Manduca sexta* and phylogenetically related sphingids (Lepidoptera : Sphingidae). *Chemoecology* **12**, 29–46.

Wright, S. H. and Dantzler, W. H. (2004). Molecular and cellular physiology of renal organic cation and anion transport. *Physiol. Rev.* **84**, 987–1049.

Wu, D. S. and Beyenbach, K. W. (2003). The dependence of electrical transport pathways in Malpighian tubules on ATP. *J. Exp. Biol.* **206**, 233–243.

Xu, W. and Marshall, A. T. (1999a). Magnesium secretion by the distal segment of the Malpighian tubules of the black field cricket *Teleogryllus oceanicus*. *J. Insect Physiol.* **45**, 777–784.

Xu, W. and Marshall, A. T. (1999b). X-ray microanalysis of the Malpighian tubules of the black field cricket *Teleogryllus oceanicus*: the roles of Na K-ATPase and the Na K 2Cl cotransporter. *J. Insect Physiol.* **45**, 885–893.

Yamamura, M., Suzuki, K. T., Hatakeyama, S. and Kubota, K. (1983). Tolerance to cadmium and cadmium-binding proteins induced in the midge larva, *Chironomus yoshimatsui* (Diptera, Chironomidae). *Comp. Biochem. Physiol.* **C75**, 21–24.

Yang, J., McCart, C., Woods, D. J., Terhzaz, S., Greenwood, K. G., ffrench-Constant, R. H. and Dow, J. A. (2007). A *Drosophila* systems approach to xenobiotic metabolism. *Physiol. Genomics* **30**, 223–231.

Yoshida, N., Koizumi, M., Adachi, I. and Kawakami, J. (2006). Inhibition of P-glycoprotein-mediated transport by terpenoids contained in herbal medicines and natural products. *Food Chem. Toxicol.* **44**, 2033–2039.

Yu, M.-J. and Beyenbach, K. W. (2001). Leucokinin and the modulation of the shunt pathway in Malpighian tubules. *J. Insect Physiol.* **47**, 263–276.

Yu, M-J. and Beyenbach, K. W. (2004). Effects of leucokinin-VIII on *Aedes* Malpighian tubule segments lacking stellate cells. *J. Exp. Biol.* **207**, 519–526.

Zdobnov, E. M. and Bork, P. (2007). Quantification of insect genome divergence. *Trends Genet.* **23**, 16–20.

Zhang, S. L., Leyssens, A., Van Kerkhove, E., Weltens, R., Van Driessche, W. and Steels, P. (1994). Electrophysiological evidence for the presence of an apical H^+-ATPase in Malpighian tubules of *Formica polyctena*: intracellular and luminal pH measurements. *Pflug. Archiv.* **426**, 288–295.

Zhao, Y., Meredith, J., Brock, H. W. and Phillips, J. E. (2005). Mutational analysis of the N-terminus in *Schistocerca gregaria* ion-transport peptide expressed in *Drosophila* Kc1 cells. *Archiv. Insect Biochem. Physiol.* **58**, 27–38.

Zhou, G., Kohlhepp, P., Geiser, D., Frasquillo, M. D., Vazquez-Moreno, L. and Winzerling, J. J. (2007). Fate of blood meal iron in mosquitoes. *J. Insect Physiol.* **53**, 1169–1178.

Zubrzak, P., Williams, H., Coast, G. M., Isaac, R. E., Reyes-Rangel, G., Juaristi, E., Zabrocki, J. and Nachman, R. J. (2007). Beta-amino acid analogs of an insect neuropeptide feature potent bioactivity and resistance to peptidase hydrolysis. *Biopolymers* **88**, 76–82.

Collective Decision-Making and Foraging Patterns in Ants and Honeybees

Claire Detrain and Jean-Louis Deneubourg

Unit of Social Ecology, Université Libre de Bruxelles CP231, Campus de la Plaine, Brussels, Belgium

1 Introduction

Organisms inhabit a heterogeneous world in which resources are distributed patchily, vary in size or quality over time, and trigger the interest of several competing species. In addition, the physical properties of

ADVANCES IN INSECT PHYSIOLOGY VOL. 35
ISBN 978-0-12-374329-9
DOI: 10.1016/S0065-2806(08)00002-7

the environment – such as the air wind-flow for flying species or the substrate granularity for walking ones – determine the energetic costs, as well as the time investment associated with the searching and the exploitation of food resources. Hence, during its entire life, each individual forager has to choose from a range of alternatives: it has to decide not only which food type to exploit, but also which part of its home-range to explore, which foraging risk to face or which path to follow. Decision-making is even more complex for animals living in social systems (Krause and Ruxton, 2002). Indeed, the decision of each individual is closely related to the behavior of the other group members. Such social influence on individual decision-making is essential to group-living animals, for which cooperation plays a key role by enhancing fitness. It is a prerequisite for the emergence of coordinated behavior; for example, the reaching of a consensus about moving in a common direction, or the collective choice of exploiting a single food patch. Situations in which group members choose between two or more mutually exclusive actions with the specific aim of reaching a consensus are usually termed as "collective decision-making" (Deneubourg and Goss, 1989; Conradt and Roper, 2005; Couzin *et al.*, 2005; Sumpter, 2006; Dyer *et al.*, 2008). Examples of collective decisions were demonstrated in several different taxa and contexts such as the choice of a common feeding place in birds (see e.g. Perusse and Lefebvre, 1985) or the choice of a common moving direction in fish schools (e.g. Couzin *et al.*, 2005) and in mammal troops (e.g. Kummer, 1968; Boinski and Garber, 2000). Put in a historical perspective, the study of collective decision-making find its roots in political sciences and several theories have been built since Condorcet's (1785) formulation of the voting paradox. These theories have shown how individual preferences may lead to quite different social choices depending on the underlying voting procedures (Gaertner, 2001). Even though political sciences and life sciences may seek to answer similar questions – and often share the same vocabulary (List, 2004) – when they talk about decision-making, collective choices made by animals are, however, governed by procedures that are very different from voting procedures in humans.

Ants and honeybees are useful biological models for studying collective decision-making processes in group-living animals. Insect societies containing up to several thousands of individuals nevertheless succeed in behaving in a coordinated way. Both the ant nest and the beehive display striking examples of collective responses in a variety of contexts such as foraging (see references further in the text), nest moving (Franks *et al.*, 2002; Pratt *et al.*, 2002; Seeley and Visscher, 2004; Pratt and Sumpter, 2006; Seeley *et al.*, 2006) or territorial defense (Lumsden and Hölldobler, 1983; Detrain and Pasteels, 1992; Millor *et al.*, 1999). In most cases, the best food patch, the best nesting site or the safest place is selected. Amazingly, hundreds of nestmates, each with their own agenda, their own past experience, as well as their own behavioral profile, are able to coordinate their efforts to behave in a highly cohesive way. Beehives or ant nests then appear as integrated social structures that gather a vast array of information acquired by its members, and that compare the relative value of each resource in relation to actual colony needs. These societies "wisely" deploy their workers over their home-range area. Moreover, collective decisions give rise to social-built structures that are characterized by relatively larger spatio-temporal scales than those observed at the individual level (Camazine *et al.*, 2001). Indeed, while one ant individual can only forage over a few square meters at most, the trail network of the whole colony can extend over several acres. Likewise, while one forager, due to physiological limitations (e.g. water loss), can only stay outside the nest for, at most, a couple of hours, the colony taken as a whole can forage over food patches for relatively longer time scales – i.e. for several days or weeks.

In the past decade, life scientists have faced the exciting challenge of explaining how collective decisions may arise in societies made up of thousands of individuals. Since collective behavior may lead to a high level of cohesion, one can get the impression that "social patterns" are obeying complicated hidden blueprints or a centralized strategy imposed by one or a few leaders. This is not the case. Instead, for many social activities, the coordinated behavior of an insect colony relies on a multitude of local decisions made by individual workers. Collective structures are the dynamic by-products of interactions between colony members, and result from a decentralized decision-making process (see for example Seeley, 1995; Theraulaz and Spitz, 1997; Camazine *et al.*, 2001; Detrain and Deneubourg, 2006; Sumpter, 2006). Any individual that makes a decision based on its local perception of the environment may contribute to the final foraging decision of the whole colony. Since each individual may influence the colony's response and its fitness, such an individual has to constantly monitor the environment for cues and signals that indicate whether it should change its occupation, intensify its participation in current tasks and/or transfer the acquired information to nestmates. In a foraging context, the information the individual perceives about food profitability helps each worker to decide not just how, where and when to exploit a food patch, but also whether or not to recruit nestmates.

In this review, we compare collective decisions in hives and in ant nests by relating the properties of recruiting signals to the foraging strategies displayed by these two insect societies. We briefly describe the main positive and negative feedbacks that help foragers self-regulate their activities according to environmental constraints and opportunities. Even though honeybees and ants share a similar recruitment scenario, we would like to stress that the nature of their recruiting signals, as well as the ways information about food patches is conveyed to nestmates, differ greatly. In honeybees, tactile signals (i.e. the waggle dances) rule many aspects of recruitment, whereas chemical communication (i.e. recruitment trails) is the backbone of collective foraging in ants. We emphasize the intimate link that relates the mode of locomotion of one species (i.e. walking or flying) to the media "chosen" to communicate information about food resources. In particular, we demonstrate that the distinct properties of chemical and tactile signals match the environmental constraints faced either by ant individuals on the ground or by honeybees in the air.

One goal of this chapter is to advocate that the use of either chemical or tactile signals in recruitment has provided communication constraints, which may account for different foraging patterns displayed by ant nests or by hives. Therefore, in the second part of this review, we compare how ants and honeybees make collective choices between identical sources, code information about food characteristics and select the most profitable resources. In particular, we shall argue that differences in the collective patterns displayed by ant nests or hives depend on the level of non-linearity of their recruitment process, or, in other words, on the amplifying potential of their communication. In addition, we examine how randomness may lead to diversity and flexibility of foraging patterns, instead of simply reducing the efficiency of food exploitation.

Throughout this review, collective patterns displayed by hives and ant nests are placed in a behavioral ecology perspective: we try to connect the properties of recruiting signals to environmental constraints, as well as to the efficiency of collective foraging strategies. Finally, by referring to other group-living arthropods, we address generic issues about information processing, its dissemination among conspecifics and the emergence of cooperative behavior.

2 Building up collective foraging behavior through feedbacks

The way individuals process information about current food opportunities or foraging risks, and the way they communicate with each other are crucial to the foraging success of insect societies (Seeley, 1994, 1995; Detrain *et al.*, 1999; Sendova-Franks and Franks, 1999; Camazine *et al.*, 2001). Insect societies not only have to communicate how far and in which direction food is located, but also have to decide how many foragers should be engaged in food exploitation. Much information can be exchanged through direct interaction between nestmates, such as antennal contacts or chemical emission of pheromones. In addition, workers can obtain information through animal-induced changes within a shared environment. Indeed, any individual that modifies its environment – and hence that of its nestmates – indirectly influences the further behavior of other group members. For instance, when an individual impoverishes an area by regularly feeding on it, it decreases the foraging success rate of conspecifics and hence influences their further foraging decisions. This may lead to spatial specialization and related division of labor (Deneubourg *et al.*, 1987; Beshers and Fewell, 2001; D'Ettore, 2007; Ravary *et al.*, 2007).

Interactions between nestmates or between nestmates and their environment can be classified into two main categories: positive and negative feedback. A positive feedback is an interaction that increases or amplifies the response to an initial stimulus. One of the best-known examples of positive feedback is the emission of recruiting signals by foragers having discovered a new food source. This signal is emitted either incidentally or actively by successful scouts: it will mobilize and attract additional nestmates to the food site, which, after feeding, will in turn reinforce and amplify the recruiting signal (Hölldobler and Wilson, 1990; Seeley, 1995). However, without an antagonizing mechanism the process might become uncontrollable. Negative feedbacks, such as food exhaustion or overcrowding of food sites, act to counter or "dampen" amplifying effects that would otherwise lead to an excessive mobilization of foragers. Such mechanisms provide a stabilizing force and play a regulatory role by adjusting the number of recruited foragers to the number and size of food items. Insect societies may thus be seen as a network of coupled feedbacks. Their ecological success depends on the subtle interplay between positive and negative feedback loops that focus the colony foraging effort onto profitable sites, lead to the emergence of optimal patterns and open the way to cooperation in food exploitation and defense (see e.g. for honeybees, Seeley *et al.*, 1991, 2000; Seeley, 1995; for ants, Hölldobler and Wilson, 1990).

2.1 RECRUITMENT: THE QUINTESSENCE OF POSITIVE FEEDBACKS

Many examples have been described and a vast body of literature has been devoted to the recruitment phenomenon, notably in social insects. Recruitment can be passive, based on inadvertent cues provided by others. For instance, the feeding activity of an animal can be perceived by conspecifics that will join it at the food source. Such passive recruitment has been reported in a variety of animal species and is associated with complex cost-benefit trade-offs, since conflict between individuals often occurs when gaining access to food sources (Thomas and Valone, 1989; Valone and Templeton, 2002; Danchin *et al.*, 2004). Social insects are one example in which recruitment of nestmates is a truly active process. The successful scout "decides" to emit specific signals that attract and guide workers to the food. Hence, every time a forager returns to the nest, it brings home not only food but also information about the characteristics of the exploited resources, which it shares by means of recruitment signals. These signals provide a positive feedback that attracts nestmates, concentrates the foraging force around one food source, and contributes to a social consensus about when, where and how to forage (Camazine *et al.*, 2001).

Although both ants and honeybees are highly successful in coordinating their foraging activity and in making adaptive choices about where to forage, recruitment is achieved by using quite different communication channels.

In the case of ants, recruitment signals are essentially chemical. In some ant species, the recruiter leads one recruit (tandem recruitment) or a small group of recruits (group recruitment) from the nest to the newly discovered food source. The cohesion of the tandem or of the group is maintained by mechanical contacts between the leader's gaster and the follower's antennae and/or by short-range chemical signals emitted by the leader (Möglich *et al.*, 1974; Verhaeghe, 1982; Hölldobler and Wilson, 1990; de Biseau *et al.*, 1991, 1994; Liefke *et al.*, 2001; Franks and Richardson, 2006; Richardson *et al.*, 2007). However, in ants, the most widespread recruitment procedure is to lay a chemical trail (Fig. 1A; Wilson, 1962a,b; Beckers *et al.*, 1989; Traniello and Robson, 1995). In this case, the odor trail laid by the successful scout on its way back to the nest triggers the exit of nestmates and guides them to the food source. In turn, the recruits, having found and fed at the food source, reinforce the trail when returning to the nest. In doing so, successful foragers draw out of the nest an increasing number of nestmates and provide them with an efficient guiding trail.

In the case of honeybees, several hundred foragers bring in information about food sources and maintain a strong flow of information about food opportunities to the hive. The honeybees have a unique way of recruiting nestmates to food sources; the famous waggle dance (Fig. 1B) (von Frisch, 1967). Waggle dances, which consist of a series of waggle runs each followed by a semi-circular return, convey information about the direction and distance of the foraged site visited by the dancing bee. The angular deviation of waggle runs from the vertical corresponds to the angle of the food source relative to the current sun azimuth. The distance between the nest and the feeding site is encoded in the duration of the waggle run. Dances are thus a sophisticated means for honeybees to translate into a coded language the food location (for a review see Dyer and Seeley, 1991; Seeley, 1995; Dyer, 2002). The information encoded in this symbolic language is then read by unoccupied foragers and used to fly off and locate the foraging site. In addition to acquiring distance and direction information, bees that follow dances also learn the odor of the food and will then search for that scent in the field (Gould, 1976; Sherman and Visscher, 2002).

2.2 THE WAY YOU MOVE RULES THE WAY YOU RECRUIT

One may easily conceive that the way in which an animal guides recruits to food sources is constrained by the way the animal moves. Walking on the

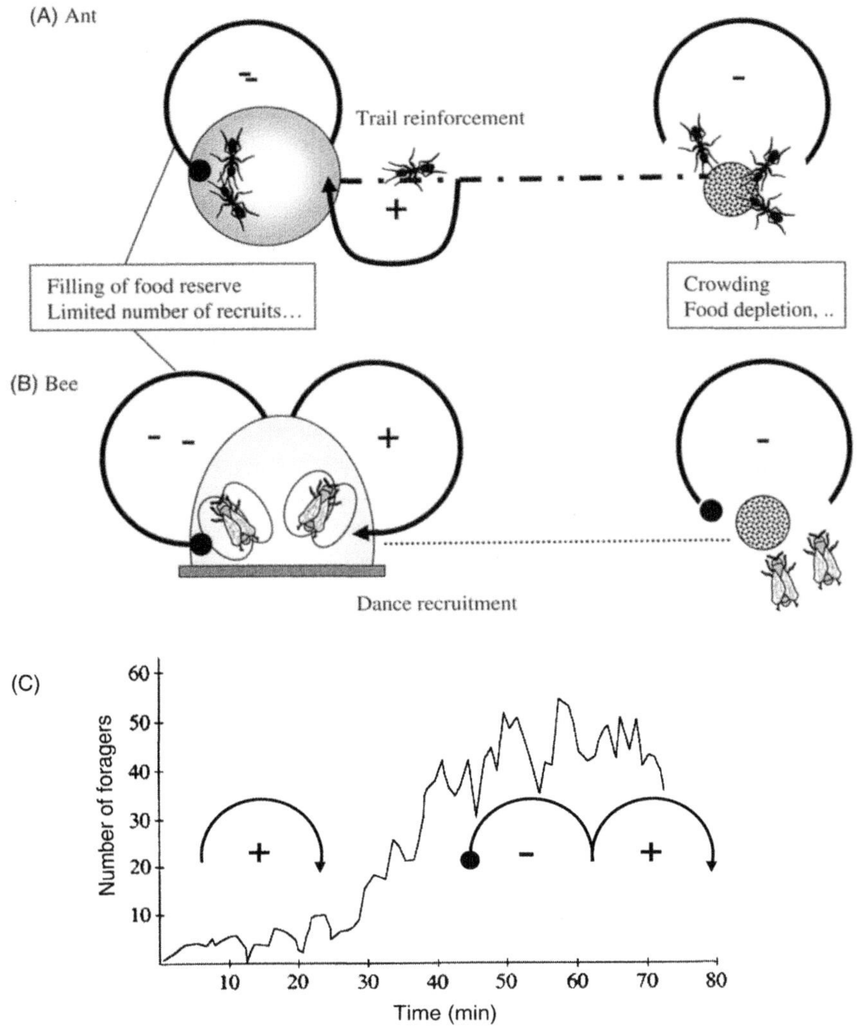

FIG. 1 Coupled feedbacks loops and dynamics of foraging in ants (A) and honeybees (B). Positive feedbacks such as ant trail reinforcement or honeybee dances trigger the exit of recruits. Thereafter, recruitment slows down due to the increasing impact of negative feedbacks, which take place inside the hive/the ant colony as well as around food sources. As a result, the dynamics of growth of numbers of foragers follows a logistic curve [(C) experimental curve for *Lasius niger* ant colonies exploiting a 1 M sucrose solution (Camazine *et al.*, 2001)].

ground or flying through the air has presumably imposed strong constraints on the type of recruitment signals used by ants and honeybees.

Quite obviously, the way ant scouts recruit nestmates is intimately related to their ability to lay signposts while walking between the nest and the food source. Successful foragers returning to the nest can easily bend their gaster downwards and hence exert a step-by-step control of the amount of recruitment signals laid over the substrate. The chemical trail, which is progressively built and reinforced by successful foragers, is then followed by ants as an invisible highway on the ground connecting the ant nest to the food site. In this case, information about the location of food, such as its spatial coordinates, is not explicitly coded by the trail or by any other signals emitted by recruiters inside the nest. In other words, recruits that exit the nest and start following the trail, have no idea where the food is located or at which distance from the nest. Foragers track the trail at a local scale through osmotropotaxis by moving their antennae over the substrate until they reach the food source (Hangartner, 1967, 1969; Detrain *et al.*, 1988; Hölldobler and Wilson, 1990; Calenbuhr and Deneubourg, 1992; Calenbuhr *et al.*, 1992). The recruitment trail is like an Ariadne's thread, guiding foragers to a food source and providing them with step-by-step directional information. The use of a trail is not limited to ants, since other walking social insects, such as termites, rely on this mode of communication as well. Indeed, trail pheromones elicit recruitment and orientation for termite species in which workers forage outside the nest or move from one nesting site to another (Kaib *et al.*, 1982; Traniello, 1982; Heidecker and Leuthold, 1984; Reinhard and Kaib, 1995, 2001; Pasteels and Bordereau, 1998; Traniello and Leuthold, 2000).

In addition to its guiding role, the following of a trail also gives an opportunity for foragers to come across many nestmates. During these head-on encounters on foraging trails, ant workers perceive several cues informing them not only about the partner's identity but also about the type of food being exploited or about the profitability of a foraging patch (Roces, 1990; Roces and Nunez, 1993; Howard *et al.*, 1996; Le Breton and Fourcassié, 2004; Dussutour *et al.*, 2007a,b). For instance, contact with food residues over the recruiter's body informs recruits about the food characteristics prior to their arrival at the feeding site and activates a food searching pattern leading to an increase in patch finding success (Le Breton and Fourcassié, 2004). Conversely, flying insects, such as honeybees, cannot exchange information about food sources along their foraging journey: they have to estimate food characteristics within the hive – namely through the perception of floral odors before flying out to the flower patch (Seeley, 1995).

Whereas the laying of chemical markers on the ground is a reliable way for walking insects to guide nestmates, it is far less convenient for flying animals, such as bees. Actually, the use of chemical trails by flying insects is limited to a few species of social stingless bees, which show a range of behavior; from odor-marking of the food alone to full trail-marking between the nest and the food source (Nieh *et al.*, 2003). In these melipone species (Lindauer and Kerr, 1958, 1960; Michener, 1974; Jarau *et al.*, 2003,

2004; Schmidt *et al.*, 2006a,b; Schorkopf *et al.*, 2007), scent-marking bees rub their proboscis over the substrate, start laying labial gland secretions and land regularly to deposit marks over twigs and leaves on their way back to the nest. The recruiting bee then triggers the exit of nestmates, which are guided by the scent trail towards the food source. The shorter foraging distances, as well as the larger sizes of stingless bee colonies compared to those of honeybees, have presumably facilitated the use of such chemical trails (Lindauer and Kerr, 1958; Beckers *et al.*, 1989; Jarau *et al.*, 2003, 2004; Schmidt *et al.*, 2006a). Conversely, for honeybees, flying over long distances has set several constraints that impede the use of odor trails. First, chemical marking of the substrate requires frequent landings and takeoffs, which may become costly in time and energy for the bee scout over long distances. In addition, aerial trails emitted by a flying scout would be subjected to wind or air turbulence and too labile to be an accurate and reliable orientation cue for recruits over long distances. These flight-related constraints have prevented the building and/or the efficient following of chemical trails by honeybees. By developing their unique mode of communication about food resources through dances, honeybees have freed themselves from flight-related constraints on information transfer. The spatial information encoded within the dance allows recruits to be immediately informed about how to reach the food (Riley *et al.*, 2005), while the floral odor inside the nest (Sherman and Visscher, 2002; Farina *et al.*, 2005; Diaz *et al.*, 2007) allows recruits to pinpoint the resource in the final stages of their flight as they approach the flower patch. Instead of being progressively built up through successive trail reinforcements (as in ants), information about the location of flower patches is "summarized" inside the hive by the honeybee dancers. While ant recruits walk along the trail without knowing their final destination, dance followers know the direction and the distance of flower patches before departing from the hive. In the hive, waggling runs along the comb provide an updated and accurate "flying map" to nestmates, which respond immediately with a direct flight to the specified patch.

2.3 REGULATION AND NEGATIVE FEEDBACKS

The ecological success of social insects depends not only on their ability to recruit nestmates but also to adjust the number of recruits to the available resources. To achieve such regulation of the workforce, positive feedbacks such as recruitment signals have to be coupled with negative feedbacks. The latter are often by-products of the foraging activity, such as crowding effects, food exhaustion or filling of nest reserves (Figs. 1A and 1B; de Vries and Biesmeijer, 1998, 2002; Dussutour *et al.*, 2004; Schmidt *et al.*, 2006b). As a result, the growth of the number of foragers at a food source can usually be defined by a logistic curve, which increases exponentially until negative feedbacks create a plateau. These negative feedbacks include the limited number of potential recruits within the nest or limited access to the food source (Fig. 1C).

In ant populations, food exhaustion or overcrowding is known to weaken or even suppress recruitment (Fig. 1A). This adaptive response is based on a simple rule evidenced by Wilson (1962a,b); a recruited ant having no more access to food goes back to the nest without laying a trail. As the trail is no longer reinforced, the pheromone evaporates and recruitment comes to an end (Wilson, 1962a,b). Such an "all-or-nothing" response of foragers to food access is explained by a more generic rule of thumb: the "desired volume" rule. A forager having found a food droplet will lay a trail only when it is able to ingest a desired volume of its own that acts as a threshold value (Mailleux *et al.*, 2000, 2003a, 2005, 2006). The use of such a criterion by ants is not restricted to a single source but can be generalized to multiple food sources, such as in the case of aphid honeydew droplets. This behavioral rule is a powerful and economical means for the society to adjust the number of potential food-carriers to the global quantity of available honeydew (Mailleux *et al.*, 2003b). When an aphid colony is overexploited, aphid-tending ants can no longer ingest their desired volume, and so stop reinforcing the trail and recruitment slows down due to pheromone decay. Such negative feedback allows a constant readjustment of the foraging force to the current productivity of aphid colonies. At any time, information about the food volume still available is conveyed to the society by the number of trail-laying ants – i.e., individuals that have reached their desired volume. In some species, a volatile, no-entry pheromone may also aid in trail choice during active foraging by closing off unrewarding paths at the bifurcations of the trail network (Stickland *et al.*, 1999; Robinson *et al.*, 2005). The coupling between such negative feedback and amplifying recruitment signals may lead to dynamics that are more complicated than simple sigmoidal ones (Fig. 1C), such as oscillations in the number of feeding foragers (Verhaeghe and Deneubourg, 1983) or periodic activity patterns (Rissing and Wheeler, 1976; Franks and Fletcher, 1983; Goss and Deneubourg, 1989). In addition, any environmental constraint that alters the dynamics of trail recruitment can contribute to the regulation of foraging without any explicit coding of information by the worker individuals (Dussutour *et al.*, 2006). For instance, if an ant colony is given the choice between two paths at low densities, it will forage over one trail but, in crowded conditions, both branches will be equally followed. Indeed, high rates of collision between ants leaving the nest and those returning with food force some of them to take the second path. This negative feedback, the collision rate, guarantees that an optimal rate of food return is maintained (Dussutour *et al.*, 2004) and also generates a temporal organization of foragers' traffic (Dussutour *et al.*, 2005b). Negative feedbacks may also occur within the ant nest (Fig. 1A). For example, when faced with different food sources (e.g. proteins versus sugars), workers will selectively collect the food type that best meets the colony's requirements. In this case, we suspect that the propensity of inner nest workers to accept trophallaxis or to respond to recruiting stimuli provides a cue to the recruiter about the actual colony food demand, as well as about the level of nest reserves (Howard and Tschinkel, 1981; Sørensen *et al.*, 1985; Camazine *et al.*, 1998; Cassill and Tschinkel, 1999a,b; Cassill, 2003). Any difficulty experienced by a recruiter in unloading its crop content may indicate a decreased interest for such a resource and lead to its progressive neglect by foragers.

In honeybees, the mechanisms that regulate recruitment to flower patches exhibit strong similarities with those of ants (Fig. 1B). Several negative feedback loops may dampen the recruiting response triggered by dancing foragers. One should notice that, unlike in ants, most of the negative feedbacks discovered so far take place inside the hive. Indeed, a bee that returns to the nest with a load of nectar has to find a receiver bee before recruiting and going back to the flower patch. The search time experienced by this forager, before being relieved of its crop load, depends on the number of bees that are simultaneously free and is also highly correlated with receiver motivation to accept such a food resource (Seeley and Tovey, 1994). Short latencies before being unloaded increase the probability of a forager carrying out a large number of waggle dances. Conversely, long latencies result in foragers queuing up to be unloaded and performing fewer waggling runs or none at all. In some cases, the unloaded bee may start a tremble dance, which fosters additional bees to function as receiver bees (Seeley, 1992). Even though tremble dancing may somewhat prolong food retrieval, foraging progressively slows down, as no more receiver workers are available within the hive (Seeley *et al.*, 1996). The decision of a bee scout to recruit nestmates is thus based on a simple cue: unloading time, which acts as a feedback mechanism to reduce recruitment effort towards resources of decreasing interest to the hive. This decision rule is a reliable method to assess the demand–offer balance for a given resource: it integrates information about the number of empty cells free for food storing in the comb, the food demand of larvae, as well as the number of nestmates available for food processing in the hive. Other feedback loops may also occur outside of the hive and take place at the flower patch itself. Because the patch size is limited, some bees may have to fly around and wait for a place at the source to become free or return empty (de Vries and Biesmeijer, 1998, 2002). Chemical cues, such as marking a deteriorated source with a repellent pheromone, may also be used by bees to slow down foraging recruitment and prevent further visits by foragers (Giurfa, 1993).

In insect societies, the amplifying effects of recruitment are thus counteracted by negative feedbacks, the importance of which increases as food sources become depleted, overcrowded or of lesser interest to the colony. Even though hives and ant nests display a similar logic in foraging regulation, the nature of feedback loops involved is quite different. Besides, as discussed in the following sections, the different ways foragers are recruited lead, at the collective level, to quite different collective choices and foraging patterns that are non-intuitive and hard to predict.

3 The way you recruit rules the way you make collective choices

Under natural conditions, ant colonies or hives may be faced with several food sources and have to make efficient collective choices between these alternatives. Many factors are likely to influence such collective choices: food characteristics may differ, competitors and predators may interfere with the colony activity, temperature and other climatic conditions may influence the energy/time costs associated with foraging. Therefore, each colony has to allocate foragers to food patches in a way that maximizes energy input and minimizes foraging costs. We review simple situations that demonstrate how competition between recruitment to several food sources may lead to collective responses that are adaptive for the whole society. First, we look at laboratory experiments with ants. Although these experiments may appear somewhat artificial, they help to identify, at the individual level, the simple recruitment rules that underlie the diverse and sometimes unexpected foraging patterns displayed by the ant colonies. Moreover, they allow us to build models and to test their predictions. Such models can be extended and adapted to the recruitment logic of honeybees (Camazine and Sneyd, 1991; Seeley *et al.*, 1991; de Vries and Biesmeijer, 1998, 2002). They are also convenient tools for comparing the impact of recruitment properties on collective decisions made by ants or by honeybees, and for discussing to what extent emergent patterns are presumably adapted to the foraging ecology of these social insects.

3.1 COLLECTIVE CHOICES AND TRAILS: EXPERIMENTAL EVIDENCE IN ANTS

The choice between identical resources is an unusual situation under natural conditions and seems of little interest from an ecological point of view. However, binary choice experiments between identical sources can be useful in highlighting how seemingly small differences in the communication process may affect collective decisions. When two identical sources are available, the most intuitive pattern would be the equipartition of foragers between the two resources. Indeed, an allocation of foragers that is strictly proportional to the availabilities of resources – i.e. an ideal free distribution of foragers between sources (Giraldeau and Caraco, 2000) – seems the best way for a species to maximize its fitness by reducing the costs of competition between conspecifics. However, quite unexpectedly, an equipartition of foragers is seldom observed in the majority of trail-laying ant species. After a short period of equal exploitation, a bifurcation is observed, meaning that one source is selected and more strongly exploited than the other, even though the two food sources are identical (Figs. 2A and 3A). This is not the result of any asymmetry of the experimental set up since the side of the chosen source may change over successive tests (Pasteels *et al.*, 1987; Beckers *et al.*, 1990; Sumpter and Beekman, 2003). Similarly, when the colony has the choice between two paths of the same length, most ants will end up traveling and sharing only one foraging route (Deneubourg *et al.*, 1990; Beckers *et al.*, 1992b; Dussutour *et al.*, 2004). The process of choice at the colony-level, between equivalent branches or between identical food sources, is thought to start by chance with an imbalance in the pheromone amounts between the two branches, large enough to trigger a positive feedback loop. The disproportional numbers of workers choosing and laying pheromones on the more traveled branch means that the ants' stream to one source/path out-competes the stream to the other source/path (Goss *et al.*, 1989).

3.2 A MODEL OF FOOD SELECTION INVOLVING CHEMICAL TRAILS

The bifurcation process – i.e. the selection of one food source among several identical ones – can be predicted by a behavioral model of food recruitment that relates the behavior of recruiting individuals to the foraging choice emerging at the society level (Deneubourg and Goss, 1989; Beckers *et al.*, 1992a, 1993; Nicolis and Deneubourg, 1999; for a review see e.g. Detrain *et al.*, 1999; Sumpter and Beekman, 2003; Sumpter and Pratt, 2003). This model, built upon empirical findings, describes how the amount of trail pheromone, and hence the traffic of ants on each trail, will change over time. In this model, an ant leaves the nest, chooses a path, reaches one of the two food source, ingests food and promptly returns to the nest, laying a trail pheromone. Each ant has to make two decisions: (1) to leave the nest or not, and (2) to choose one of the trails leading to equidistant food sources of identical quality (i.e. a bridge with two equal branches in a binary choice experiment). For each ant, the probability of leaving the nest increases with the total amount of trail pheromone at the nest entrance. Over time, the increasing rate of nest departure coupled with trail reinforcement by successful foragers contributes to the sigmoidal growth of the recruits' population. One can neglect time variations in nest exits since

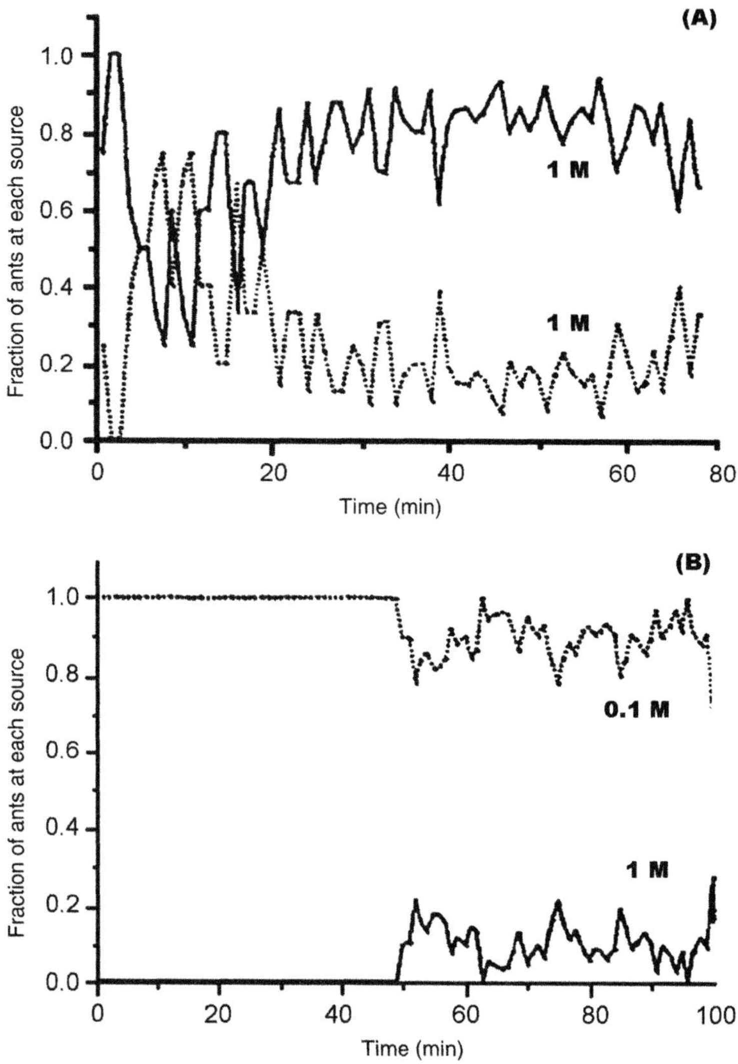

FIG. 2 Collective choice of one food source by ants faced with: two sources of equal quality (1 M sucrose solution) that are introduced simultaneously (A); two different food sources, the 1 M sucrose solution being introduced after the 0.1 M food source (B). The temporal dynamics of the fraction of ants at each food source are based on experimental data on *Lasius niger* ant colonies (Beckers *et al.*, 1990).

they do not affect the general dynamics of collective choices (Camazine *et al.*, 2001).

An ant having left the nest has to choose one of the two paths: the probability for an ant of choosing one path *i* depends on a general choice

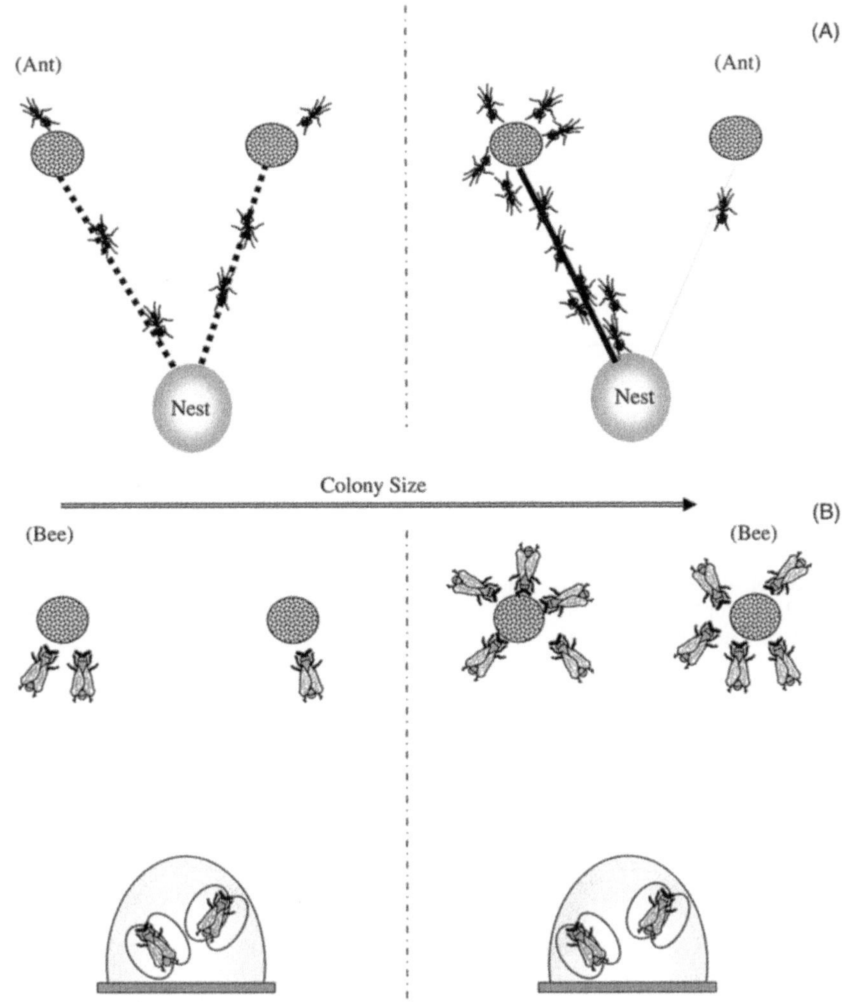

FIG. 3 Collective choice between two food sources of equal quality as a function of colony size. (A) In ants, the selection of one food source is more marked when the colony size increases. (B) Conversely, in honeybees, foragers from large-sized hives are more evenly distributed among the two food sources.

function $A_i/(\Sigma A_l)$ in which A_i is the attractiveness – i.e. amount of trail pheromone – of the trail i leading to the source i (Fig. 4). We can quantify the ant's decision at the choice point by Eq. (1) in which the orientation toward one source and/or towards one trail compared to the others

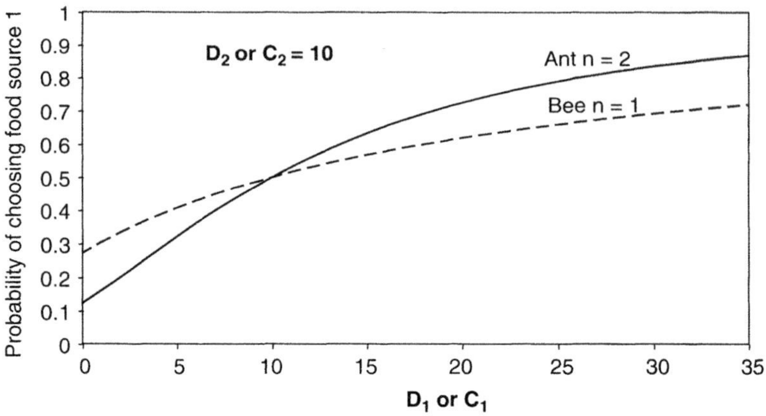

FIG. 4 Response of individual foragers to differences in recruitment signals. Solid line: Probability for one ant to choose food source 1 as a function of the trail concentration (C_1). The concentration of the trail leading to source 2 (C_2) is kept constant and equal to 10. Parameters are fixed at $n = 2$ and $k = 6$ [see Eq. (1) in the text]. Dashed line: Probability for one honeybee to choose flower patch 1 as a function of the number of bees dancing for this patch (D_1). The number of dancing bees (D_2) is kept constant and equal to 10. Parameter values are fixed at $n = 1$ and $k = 6$ [see Eq. (3) in the text].

depends on the values of the pheromone concentration c_i on each trail i (Deneubourg and Goss, 1989).

$$P_i = \frac{A_i}{\sum\limits_{l=1}^{s} A_l} = \frac{(k + c_i)^n}{\sum\limits_{l=1}^{s} (k + c_l)^n} \qquad i = 1, 2, \ldots s. \tag{1}$$

$s = 2$ in the case of a choice between two food sources.

- n determines the steepness – i.e. the level of non-linearity – of the recruitment response. A high n value means that ants are highly sensitive to changes in the intensity of recruiting signals: even a slightly higher amount of pheromone over one branch will markedly increase the probability of ants to choose this branch.
- k corresponds to the intrinsic degree of attraction of an unmarked branch of the bridge. For high k values, high differences in trail concentration between branches are needed to get a non-random choice of one path.

On its return journey from a food source, each ant lays a quantity of pheromone q_i. In the case of identical food sources ($q_1 = \ldots = q_s = q$), each ant deposits the same amount of pheromone on its homeward journey and the pheromone concentration over one trail is directly proportional to the flow of foragers (ϕ) over this trail. At each time unit, a quantity $q\phi P_i$ of pheromone is added to the trail i, while vc_i is the decay of the trail pheromone amount. The trail is thus assumed to evaporate at a rate proportional to its concentration with v being the inverse of the mean trail pheromone lifetime. The following system of s differential equations describes the rate of change in pheromone concentration on trails.

$$\frac{\mathrm{d}C_i}{\mathrm{d}t} = \phi q P_i - vC_i \quad i = 1, \ldots s \tag{2}$$

Monte-Carlo simulations based on these equations and on parameter values obtained experimentally for *Lasius niger* species give the time evolution of the ant population over each branch. Initially, both sources are exploited equally but a break suddenly occurs, after which one source becomes more exploited than the other. One food source is selected and exploited by more than 60% of foragers in nearly all cases (in around 90% of the simulations) (Fig. 5A). As for laboratory experiments (Fig. 2A), the asymmetrical exploitation of identical food sources arises as one trail becomes slightly stronger than the other, resulting in it being followed more often and further reinforced by trail-laying foragers in a positive feedback cycle. Due to the non-linearity ($n > 1$) of an ant's response to the recruitment trail, the selection of one food site arises from the amplification of slight initial differences in the relative pheromone concentration on the trails. However, one can never say which trail will dominate or which source will be chosen since this selection is based on random events.

The model can be used to assess the role of each parameter in the final selection of one food source. It reveals that collective responses can be even more complex and sophisticated. One can observe that an asymmetry in food exploitation is more marked and more quickly reached when the flow of foragers or the colony size is larger, as expected from any phenomena based on amplifying processes (Figs. 3A and 5A). Below a certain number of ants or interactions, the exploitation pattern may be symmetrical, whereas above this value a collective choice is very likely to emerge, with most of the foragers being ultimately focused on one or only a subset of available resources (Nicolis and Deneubourg, 1999). Likewise, when individual trail-laying behavior is more intense, or when the rate of pheromone decay is lower, a fluctuation in the amount of trail chemical over one path is likely to be amplified and lead towards an unequal exploitation of food sources. To observe these phenomena experimentally, one may vary the "foragers' flow" parameter by comparing colonies of different sizes, or one can increase the intensity of the trail laid by the ants (q) by increasing the quality of food sources (e.g. the sugar concentration). However, the influence of pheromone evaporation rate on the ants'

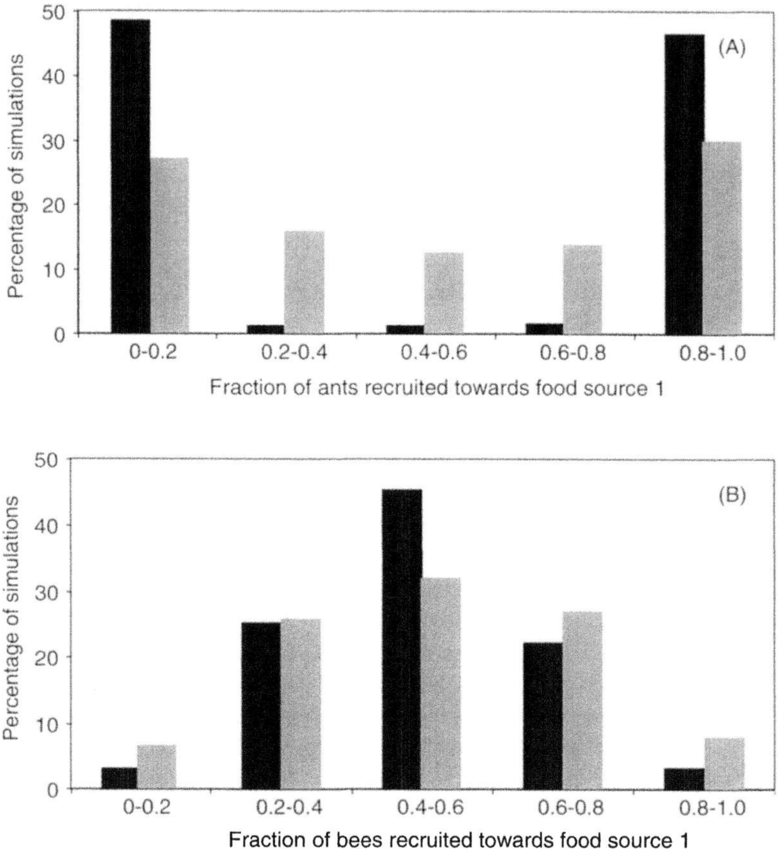

FIG. 5 Frequency distribution of simulations as a function of the fraction of foragers that choose food source 1 when two identical food sources are offered simultaneously (1000 simulations for each tested condition). (A) Distributions of ants at each food source. They were measured after 1 h and were based on the following parameter values: $k = 6$; $q_1 = q_2 = 1$; $v = 0.00056\,\mathrm{s}^{-1}$, $n = 2$ [for further explanations see Eq. (1) in text]. The bimodality of the ants' distribution is more clear-cut for high ant flows (dark bars; probability of leaving the nest (ϕ): 0.066 ant s^{-1}) than for small ant flows (grey bars; ϕ: 0.017 ant s^{-1}). (B) Distributions of honeybees at each flower patch, measured after 2 h. Simulations are a simplified version of the Camazine and Sneyd (1991) model and are based on Eq. (3) (see text). Probability of remaining a dancing bee after a foraging flight = 0.8; flying time = 250 s; resting time in the dance area = 120 s; $k = 0$; $n = 1$. Honeybees tend to be equally distributed between the two food sources with a distribution that is narrower for large forager flows (dark bars; probability of leaving the hive (ϕ): 0.05 bee s^{-1}) than for small foragers flows (grey bars; ϕ: 0.0085 bee s^{-1}).

collective pattern is more difficult to investigate experimentally since it is determined by a multitude of parameters, such as ground temperature or the perceptive physiology of the ants. In this latter case, the model has a strong heuristic potential by allowing "numerical experiments" to be performed, which would be difficult or even impossible to carry out in the laboratory.

3.3 COLLECTIVE CHOICES AND DANCES: A MODEL FOR HONEYBEES

Similar to ant colonies, the hive has to make a collective decision about which flower patch to exploit and how many foragers should be allocated to the available resources. Under natural circumstances, the concurrent availability of several food sources is probably common as plants may undergo synchronized blooming. The model developed for trail-laying ants can be transposed to food recruitment in honeybees (Camazine and Sneyd, 1991; Seeley *et al.*, 1991). The number of dancing bees is analogous to the trail concentration and determines the attractiveness A_i of one foraging site over another. In honeybees, the probability of a recruit going to an advertised site is proportional to the number of waggle runs performed by the bees recruiting for that site (Camazine and Sneyd, 1991; Seeley *et al.*, 1991). In addition, foragers dancing on the comb are able to attract with a vibratory signal dance-followers that are too far away to perceive tactile signals (Tautz *et al.*, 2001). Judd (1995) observed that, although the average number of bees orienting to a dancer can be as high as five or six, at most only one or two bees are receiving the dance information. Camazine and Sneyd (1991) have incorporated this observation in their model, so that only one bee at a time can follow the dance and extract information from it. Hence, for honey bees, the attractiveness of a food source A_i, or in other words the number of bees recruited towards food source i, is directly proportional to $(D_i d_i)^n$, where D_i is the number of bees in a dancing state. However, since these bees may stop for a while between successive dances, D_i is multiplied by the relative amount of time (d_i) during which recruiting bees actually dance. When the two food sources are of identical quality, d_1 is equal to d_2.

The probability of one honeybee foraging on flower patch i is then given by Eq. (3):

$$P_i = \frac{A_i}{\sum_{l=1}^{s} A_l} = \frac{(k + d_i D_i)^n}{\sum_{l=1}^{s} (k + d_l D_l)^n} \qquad i = 1, 2, \ldots s. \tag{3}$$

The parameter n accounts for the non-linearity of the response and reflects the sensitivity of recruits to the relative occurrence of a dance i associated with food source i. A value of $n > 1$ would mean that a recruit would be more likely to forage at a given flower patch when there are several dancers coding for this source. However, in the case of honeybees, as predicted by Camazine and Sneyd (1991) and demonstrated experimentally by Seeley *et al.* (1991), bees do not sample dancers in an effort to compare the relative profitability of food sources. Rather, a bee recruit is influenced only by a single dancing recruiter at any one moment. Since there is no synergetic effect of concurrent dances, the value of n in Eq. (3) should equal 1. Thus, the attractiveness A_i of one foraging site i over the others should be directly proportional to the occurrence of dances $D_i d_i$. As a consequence, even though the number of dances is analogous to the trail concentration, the honeybee recruitment differs from that of ants in its level of non-linearity ($n = 2$ for trail-laying ants, $n = 1$ for dancing bees). This difference has only a slight impact on the probability of one individual choosing a path or a food source (Fig. 4), but it is responsible for the very different properties of collective patterns displayed by honeybees and ants. Indeed, the model predicts that the hive will occasionally select one source only and that the most frequent response of honeybee foragers will be an even distribution between identical food sources (Figs. 3B and 5B). Furthermore, unlike ants, which more frequently select one source when the number of foragers increases (Fig. 5A), the honeybees show an even distribution of foragers whatever the hive size (Fig. 5B). This prediction finds some support in the fact that small hives foraged at approximately the same number of patches as large colonies (Beekman *et al.*, 2004). As the hive size increases, the model also predicts that the foragers will be even more equally allocated among food sites and that the presence rate of bees at one of the two food sources will shift from a flat to a mono-modal distribution with a peak at 50% (Fig. 5B). Hence, an increase in colony size has unexpected opposite effects for hives than for ant nests: larger forager flows should enhance the collective selection of one food source by ants, whereas it should draw the honeybees towards a more equal exploitation of resources (Figs. 3A and 3B). In the field, no symmetry breaking between identical sources has yet been reported for honeybees and further experimental evidence is still needed to confirm the model's predictions of an even distribution of bee foragers between flower patches. Nevertheless, one should be aware that the selection of one food source by honeybees may occur due to some "hidden" amplifying behavioral traits associated with recruitment. For instance, once a recruiter enters the hive, it stimulates forager mates to return to a known feeding place, not only by dances, but also by nectar transfers which reactivate experienced foragers. This may have a snowball effect since experienced bees follow a significantly lower number of waggle-runs and are more easily reactivated to perform nectar collection (Gil and Farina, 2002). Another source of non-linearity could be the memorization of food location, as well as the learning of flying routes that facilitate the orientation of bees and shorten their foraging journeys. These phenomena are translated, from a mathematical point of view, by an increase of the parameter n value and may lead to symmetry breakings as predicted by de Vries and Biesmeijer (2002).

In insect societies, collective decisions arise as a result of the competition between different sources of information that are amplified by recruitment signals. Through its trail-laying, an ant colony with a large forager population will select and focus its effort on one resource even if all available sources are of identical quality. Alternatively, an ant colony may exploit all sources equally if the colony itself is small or if the number of sources is very large (Nicolis and Deneubourg, 1999). Such switching in the foraging pattern, which is not predicted by the honeybee model, matches the features of food resources exploited by ants. Unlike bees that wander among short-lived flower patches, several mass-recruiting ant species feed on honeydew and forage on rather stable food sources, such as aphid colonies. By focusing the foraging force on a limited number of food sites, ant colonies can protect and monopolize these resources through cooperative defense against competitors or predators. Such an "owner strategy" can be efficiently implemented only by large sized ant nests since the costly loss of workers may be fatal to small incipient nests (Franks and Partridge, 1993; see also Mailleux *et al.*, 2003a).

By contrast, a hive presented with an array of identical sources will scarcely focus its activity on a single flower patch. It seems more advantageous for a honeybee colony to distribute thousands of workers among multiple flower patches rather than to be crowded at a single food source or to monopolize a flower patch of highly fluctuating profitability.

Hence, the different ways used by ants and honeybees to allocate foragers between identical resources are not primarily due to differences in the sensory or physiological capacities of the individuals, but are by-products of the properties of their recruiting signals (chemical trail versus dances), which shape the decision-making processes operating at the collective level.

4 Choosing the best food resource

The ant colony, as well as the hive, provides striking examples on how rather simple individual behavior and decision rules can lead to different but equally efficient patterns of food exploitation. The previous section considered the case study in which identical sources were simultaneously offered to a colony. However, the availability of food resources most often varies in space and time, while profitability differs according to several factors such as food quality, distance, safety and so on. Since individuals have only a local and partial knowledge of their environment, one may wonder how insect societies adjust their foraging effort in order to exploit preferentially the most profitable resources.

In both honeybees and ants, foragers are adequately allocated to the best food sites without any of the participants having a broad knowledge of all food opportunities. An ant or a honeybee scout does not compare nor keep track of information from multiple foraging sites: its knowledge of the array of food sources is limited to its particular food site, which it assesses using functional cues (ants: Hölldobler and Wilson, 1990; Detrain *et al.*, 1999; Detrain and Deneubourg, 2002; honeybees: Seeley *et al.*, 1991; Seeley, 1995; Biesmeijer and Ermers, 1999; Dyer, 2002; stingless bees: Schmidt *et al.*, 2006a). On the basis of this individual estimate of food profitability, the scout will decide how strongly to advertise the source by either laying a trail or by dancing in the hive. The best food sources are then selected by the whole colony – not by centralized decision-making, but simply by competition between recruitment signals differing in intensity and/or duration. While a tuning of recruitment allows both ants and honeybees to make adaptive choices, these insect societies are expected to somewhat differ in the robustness of their collective decisions. Indeed, while the removal of a dancing bee would lead to losing part of the information about food opportunities, the removal of one ant would hardly affect the amount of pheromone on the trail, which keeps the ants informed about food quality and food location.

The strength of amplification processes – i.e. recruitment rates towards different food sources – depends on two main categories of parameters (Detrain and Deneubourg, 2002, 2006). The first consists of any modulation of the recruiting signals that may enhance the mobilization of inactive nestmates towards a food source. For example, an individual scout bee may lengthen the dance duration or an ant may lay a higher amount of trail pheromone in relation to its own estimate of food profitability. The second set of factors involves various environmental features, (1) that may influence the duration of a foraging cycle such as the impact of temperature on walking/flying speed of foragers, or (2) that may alter the trail lifetime such as the absorbency of the substrate (Detrain *et al.*, 2001; Detrain and Deneubourg, 2002; Jeanson *et al.*, 2003). Likewise, a longer path toward a feeding site (Goss *et al.*, 1989) or a longer food distance from the nest (Detrain *et al.*, 1999) may reduce the rate of trail reinforcement and lead to a remote food source losing out in its competition with other resources. In these latter cases, one food source or one path can be selected without any modulation of the recruiting signal by the individuals. In the next sections, we consider only the first set of factors that act upon recruitment dynamics by reviewing how an ant or a honeybee actually fine-tunes its communication in order to select the best food sources.

4.1 FOOD SELECTION AND TUNING OF TRAIL PHEROMONE BY ANTS

In the case of trail-laying ants, if a rich source (1 M sucrose solution) is discovered at the same time as a poor one (0.1 M sucrose solution), the food source with the highest sugar content is selected and exploited by most of the foragers (Fig. 6A). There is no evidence of any qualitative change in the chemical content of recruitment trails laid by ant scouts depending on food quality. Information about food quality is "coded" only through quantitative changes in the trail intensity: more frequent trail spots will be laid

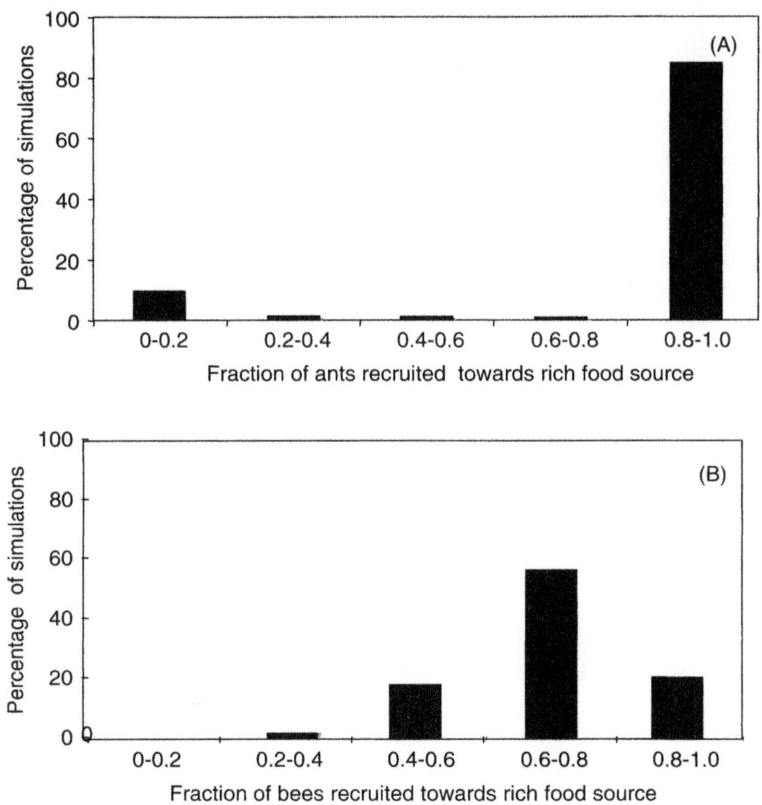

FIG. 6 Frequency distribution of simulations as a function of the fraction of foragers that choose the most profitable food between two different food sources (1000 simulations for each tested condition). (A) Ants' distributions. They are given after 1 h of recruitment for the following parameter values: $k = 6$; $n = 2$; $\upsilon = 0.00056\,\mathrm{s}^{-1}$, $\phi = 0.066\,\mathrm{ant\,s}^{-1}$; $q_1 = 1$; $q_2 = 0.7$ with q_1 and q_2 being, respectively, the trail amount laid to the rich and the poor food source. (B) Honeybees' distributions. They are given after 2 h of recruitment for the following parameter values: $k = 0$, $n = 1$; $\phi = 0.05\,\mathrm{bee\,s}^{-1}$. The relative amount of time spent in dancing for the poor flower patch versus the rich patch ($d_{\mathrm{poor}}/d_{\mathrm{rich}}$) is fixed at 0.7.

by recruiting ants to more concentrated sugary liquid sources (see e.g. Hangartner, 1969; Hölldobler and Wilson, 1990; Detrain *et al.*, 1999). Experimental data show that this trail modulation, as a function of food quality, is rather weak. For instance, in *Lasius niger*, the average number of trail marks laid towards a 1 M versus a 0.1 M sugar solution is only of 1.5 higher (Beckers *et al.*, 1992a, 1993). Then, as predicted by Eq. (1), at the trail bifurcation point, the likelihood of an individual choosing the more concentrated trail is only slightly higher than that of it choosing the less marked trail path (3–4% higher for the first steps of recruitment). The slight impact of differences in signal intensity on the choice by individual ant (Fig. 4) contrasts with the clear-cut collective choices that ultimately emerge at the colony level (Fig. 6A). Indeed, based on numerical simulations, there is always a higher percentage of foragers that orient themselves towards the sweetest source, this rich food source drawing more than 80% of ants in nearly all cases. Hence, due to the non-linearity of the response to chemical trails, even a slight modulation of the recruiting signal drives the whole colony to a nearly unanimous choice of the most profitable food source.

Besides food quality, foragers can also convey information about food size. The decision of each scout to lay a trail is ruled by simple criteria such as the ability to ingest a wanted volume of liquid food (Mailleux *et al.*, 2000, 2003b). The prey resistance to traction is another criterion used by foragers to estimate prey size: for small prey, foragers lay only a weak trail whereas, for a prey item too large to be individually retrieved, ants dash back to the nest laying a more continuous chemical trail and trigger an intense food recruitment (de Biseau and Pasteels, 1994; Detrain and Deneubourg, 1997). Such decision criteria are highly functional as they incidentally take into account multiple sources of information, such as food size, food weight or current number of ants cooperating in food retrieval (Detrain *et al.*, 1999; Detrain and Deneubourg, 2002). They also act as individual thresholds above which a scout decides to recruit nestmates so that, at the colony level, there will be a higher percentage of trail-laying foragers directed towards large food sources.

Hence, due to the autocatalytic nature of chemical recruiting signals, ants have a high potential to draw most of the foraging force towards the most profitable food source (Beckers *et al.*, 1990, 1993; Sumpter and Beekman, 2003; Portha *et al.*, 2004). Even slight changes in the trail-laying behavior of a few foragers can boost the reinforcement of one foraging path by the succeeding ants and can ultimately lead to the exploitation of the most profitable food by the whole colony.

4.2 FOOD SELECTION AND TUNING OF RECRUITMENT DANCES BY HONEYBEES

As for trail-laying ants, the recruitment activity within the hive is adjusted to both increased and decreased food profitability: honeybees will allocate more recruits to resources that have a higher pay-off for the colony. If food profitability is gauged as low by one forager, it will abandon the resource, whereas if it estimates it as high, the forager may decide to recruit nestmates and to tune the intensity of recruiting signals according to the food profitability (see e.g. von Frisch, 1967; Seeley and Towne, 1992; Seeley, 1994, 1995; Seeley *et al.*, 2000; De Marco and Farina, 2001; Afik *et al.*, 2008). Mainly quantitative changes in the recruitment signals occur. Indeed, the strength of dances inside the hive intensifies with the profitability of the flower patch. The number of waggle dances (D_i) as well as their relative duration (d_i) can then be viewed as indicators of the overall value of the nectar source, such as the distance to the flower patch or the sweetness of the nectar. A higher occurrence of dances coding for a rich resource ($D_i\, d_i$) slightly increases the likelihood of a bee being recruited towards this source (Fig. 4). This tuning of dances results in a cascade of positive feedback at the colony level so that, at the stationary state, more than 60% of the flow of foragers will be focused on the sweetest flower patch in around 80% of cases (Fig. 6B). However, in the remaining cases, the collective choice of the most profitable food source is less clear-cut than for ants, since 20–40% of honeybees go on exploiting the poorer source. These predictions have been supported by experiments showing that the hive is actually able to focus its foraging effort on the food source with the highest sugar concentration, i.e. with the highest energy reward (von Frisch, 1967; Seeley, 1995; Dyer, 2002).

4.3 REDIRECTING COLLECTIVE CHOICE IN A CHANGING ENVIRONMENT

Ants and honeybees share the same potential to select the best food source when several resources are discovered simultaneously (Fig. 6). However, hives and ant nests differ greatly in their ability to redirect their choice depending on the temporal sequence of food discoveries.

In the case of ants, if a rich source (1 M sugar solution) is discovered *after* recruitment is well underway to a first discovered but poorer source (0.1 M), then the poor source will remain the most exploited while the rich one will be neglected (Figs. 2B and 7). Theoretical simulations predict that the poor source will keep attracting the majority of foragers in all cases (Fig. 8A). The difficulty ant colonies have in redirecting their foraging activity towards more profitable sources has been experimentally confirmed in several mass-recruiting species (Beckers *et al.*, 1990; Traniello and Robson, 1995). Similar to ants, scent-marking stingless bees may be "trapped" in the exploitation of the first discovered food source (Schmidt *et al.*, 2006a). When stingless bees experience an increase of sugar concentration over time, the newcomers at the richer source never outnumber the bees that have already exploited the poor source and reinforced the scent trail (Biesmeijer and Ermers, 1999). The difficulty of both ants and stingless bees to redirect foraging towards more profitable food sources could be

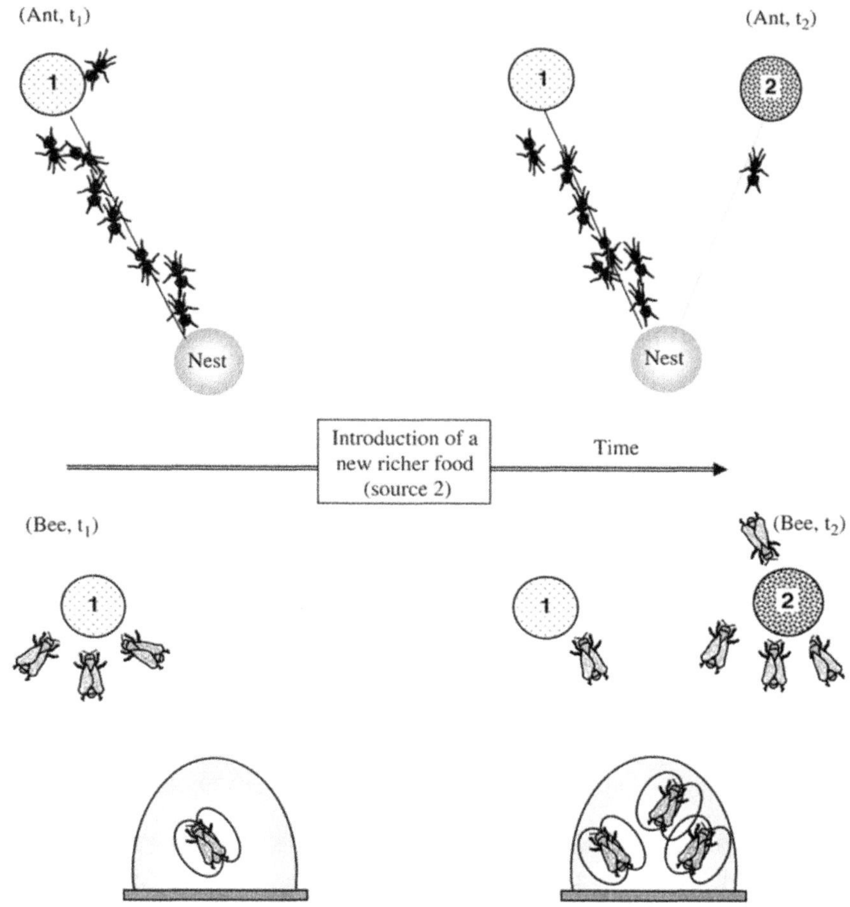

FIG. 7 Flexibility of foraging choices in ants and honeybees. In ants, the first exploited food (source 1 at t_1) will remain preferentially foraged at t_2 even a richer food (source 2) has been introduced. Conversely, honeybees that exploit the first discovered food (source 1 at t_1) are able to shift their activity at t_2 towards a more profitable resource (source 2) that has become available.

due to the chemical nature of their communication. Indeed, chemical recruiting signals elicit responses, which are strongly non-linear and highly sensitive to slight differences in signal intensities. The new trail to the richer source will never succeed in overcoming the ongoing recruitment to the first discovered food source despite its lower profitability. Likewise, the selection rate of a food source is correlated to the number of workers

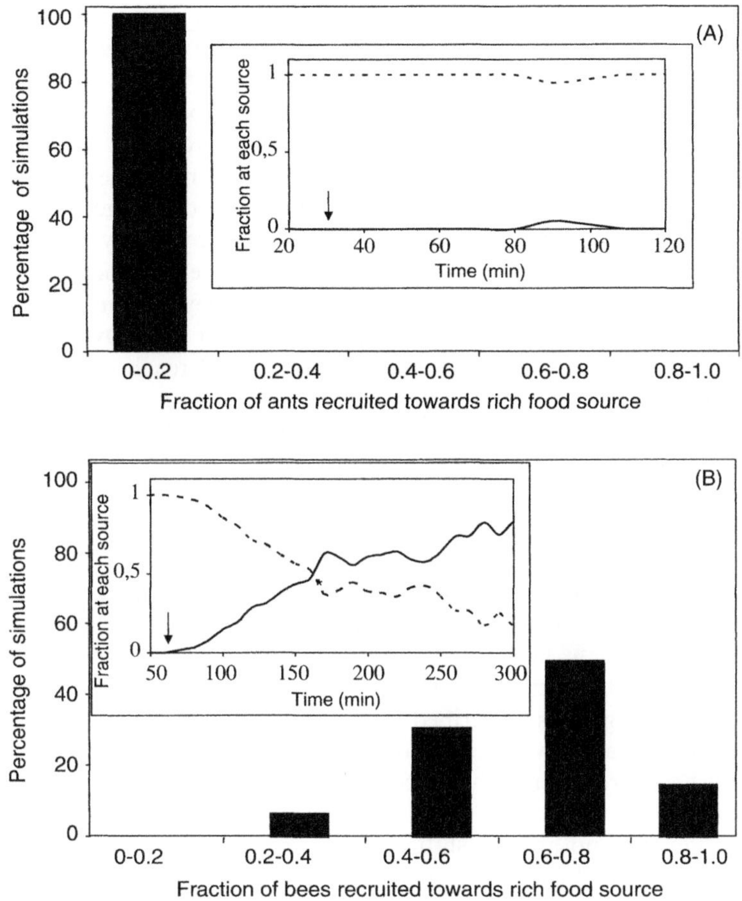

FIG. 8 Collective response to the delayed introduction of a new richer food source in ants and honeybees. (A) Frequency distribution of simulations as a function of the fraction of ants exploiting the rich source (after 120 min; 1000 simulations). When the recruitment is well-established to the first poor source, a richer source is introduced at $t = 30$ min (arrow). Even at 120 min, the poor source remains selected in all cases. (A, inside box) A simulation showing the dynamics of the fraction of foragers at the poor (dotted line) and at the rich (solid line) food source. Only a few individuals were occasionally observed at the rich source. (B) Frequency distribution of simulations as a function of the fraction of bee foragers exploiting the rich source (after 360 min, 1000 simulations). In honeybees, due to the slow growth of recruitment, foragers have access to the first poor source for a longer time before a new richer source is introduced at $t = 60$ min (arrow). At 360 min, the rich source is exploited by more than 60% of the honeybee foragers in nearly all cases. (B, inside box) A simulation showing the progressive shift of bees from the first flower patch (dotted line) towards the more profitable patch (solid line).

already exploiting it and this is likely to increasingly reinforce the scent trail (Pasteels *et al.*, 1987; Goss *et al.*, 1989; Sumpter and Beekman, 2003; Sumpter and Pratt, 2003).

Conversely, in the case of honeybees, the hive is expected to be able to select the most profitable nectar source, even if poorer food sources have been previously discovered. If between two nectar sources, the sugar concentration of one suddenly increases, the number of foragers exploiting the other source decreases, as experimentally shown by Seeley *et al.* (1991). As predicted by theoretical simulations, the delayed introduction of a sweeter source induces a reorientation of the honeybee foraging at the expense of the first discovered source (Figs. 7 and 8B). Over time, the sweetest food source will attract more foragers, will reach the same level of exploitation as the first food source and ultimately will become preferentially foraged by honeybees. In the majority of cases, 60–80% of bee foragers will end up redirecting and focusing their activity towards the rich food source (Fig. 8B). The high flexibility of honeybee foraging patterns is due to the proportional relation between occurrence/duration of dances and the number of recruited foragers. Provided that inactive workers are still available within the hive and ready to engage themselves in foraging, they can be recruited towards a richer flower patch by dancers coding for this new target. The reorientation towards more profitable resources may be facilitated by the phenomenon of cross-inhibition between forager groups (described by Seeley, 1995). Indeed, upon entering the hive, bees loaded with nectar of lesser value experience longer search times for nectar receivers, which decrease the rate of dances coding for this poor source.

Being flexible and able to shift easily to a more rewarding food patch provides obvious benefits to the insect society in terms of optimal exploitation of environmental resources. One may question why ants differ from bees by displaying such poor flexibility in their foraging patterns. The answer depends in part upon whether there is a selective advantage in being aggregated to concentrate the colony's efforts on a single site, or whether it is better to distribute colony workforce more widely. While the amount of flower nectar collected by the hive results from the simple summation of foraging journeys by crop-loaded bees, the food amount retrieved to the ant nest is more variable. Indeed, it closely depends on the ability of workers to cooperate in the killing of prey or to collectively retrieve large food items. Besides, while cooperative defense in bees is restrained to the hive surroundings (Guzmannovoa and Page, 1994; Millor *et al.*, 1999; Hunt *et al.*, 2003), ants benefit from cooperative defense of food sites and their aggressiveness toward aliens increases with the number of surrounding nestmates (see e.g. Pontin, 1961; Levings and Traniello, 1981; Lumsden and Hölldobler, 1983; Czechowski, 1984; Detrain and Pasteels, 1992; Sakata and Katayama, 2001). Concentrating the foraging effort on a few patches gives the ant colony an edge when faced with competitors, and limits the investment in defending food sources with the highest rates of return. Moreover, within a large group of cooperating ant workers, specialized castes and task-partitioning become possible, which may contribute to an increase in global foraging efficiency (Detrain and Pasteels, 1991; Ratnieks and Anderson, 1999; Anderson *et al.*, 2002).

As regards honeybees, due to dance recruitment, scouts having discovered more profitable flower patches can easily redirect the hive activity towards these new resources. The flexibility of the dance recruitment, coupled to an intense exploration, allows the hive to maintain a long-term foraging success even when flowering patches are ephemeral and variable in richness. Dance-language recruitment is presumably well-adapted to the resource variability experienced by the hive (Seeley and Visscher, 1985; Sherman and Visscher, 2002).

All these examples have questioned whether and how insect societies make collective decisions in order to exploit the most profitable food sites. There are striking similarities in the mechanisms allowing ants and honeybees to reach such a collective decision: no individual visits all food sources and none compares them nor decides which one is the best. Instead, accurate collective decisions arise from the competition between positive feedback loops and from the tuning – even if slight – of recruiting signals towards the various food alternatives. The different pools of foragers "compete" to recruit more of their comrades towards each food location. However, the amplifying potential of recruitment strongly differs between ants and honeybees with a higher level of non-linearity associated with trail pheromone signals than with dances. Such a difference has a deep impact on the properties of patterns emerging at collective level. The high non-linearity of chemical trails will make the occurrence of bifurcation more likely, i.e. collective choice of one food source, when identical food sources are offered simultaneously. On the other hand, the non-linearity of the response to chemical recruitment in ants impedes the reallocation of foragers when food profitability changes.

5 Randomness and foraging: adaptive tuning of error in ants and honeybees

In the previous sections, the recruitment process was described in an idealized way so that every recruited worker succeeds in reaching the food target. However, for a long time, biologists have been puzzled by the following paradox: impressive structures and consensus emerge at the society level, while individuals show random, highly variable and, in some cases, not very efficient behavior [see e.g. Grassé stimergy theory (1959) about the building behavior of termites]. The duality between the stochasticity of individual behavior and the determinism of the colony response is not peculiar to insect societies and occurs in other biological and non-biological systems (Nicolis and Prigogine, 1977). To solve this paradox, many researchers have argued that social insect behavior is less erratic than it appears, or have considered randomness as an unavoidable side-effect of animal behavior. Insect societies are expected to be able to afford more easily such a variance because of the large number of interacting workers (Oster and Wilson, 1978). If one ant fails, another will succeed: some inefficiency due to randomness in individual behavior is compensated by the reliability of the whole system, which will ultimately reach the goal. However, stochastic events should not be reduced to side-effects since they may have more creative and adaptive consequences (Deneubourg *et al.*, 1983). At first sight, the most accurate system of communication may seem the best one, and this is certainly the case when only one food source is present and located at the same place for a while. However, natural environments are more complex, with several sources that may fluctuate or be simultaneously available to the ants. The "lost" recruits that wander around the foraged sites may then discover new food sources of higher profitability. Hence, when multiple food sites are available, there is an optimal error level which minimizes the time needed for discovering food sources and which maximizes foraging efficiency.

One may question how an insect society finds the optimal strategy between a high accuracy of communication allowing immediate food exploitation and a low accuracy allowing new food discoveries. In ants, randomness is intimately related to the ability of foragers to orient and to follow a chemical trail. Recruits that leave the nest may lose the trail and never reach the food source. Hangartner (1967, 1969) showed that orientation along a trail is mainly due to osmotropotaxis, meaning that the ant perceives different concentrations in trail pheromone between its left and right antennae. For a given concentration in trail pheromone, the number of ants that are still present on the trail decreases exponentially as a function of the length of trail already followed (Pasteels *et al.*, 1986; Detrain *et al.*, 1988; Calenbuhr and Deneubourg, 1992; Calenbuhr *et al.*, 1992). The mean length of trail followed by ants increases with trail concentration, and the ants' orientation may become impaired above a certain concentration of trail. Since the amount of trail pheromone determines the width of its active space, it determines the probability of foragers staying within the pheromonal tunnel during their sinuous walk to reach the food source. Hence, an ant colony can easily modulate the accuracy of recruitment, by varying the quantity of trail pheromone deposited. If the food source is of high quality, a strong trail and its large active space will make its loss less likely by foragers and will guide nearly all recruits to the feeding site. If the food source is of poor quality, a weak trail may result into more frequent exits from the active space, may lead to a more diffuse spatial distribution of workers and may promote an extensive search for new food sources over the foraging area (Deneubourg *et al.*, 1983; Edelstein-Keshet *et al.*, 1995). Similarly, at the beginning of recruitment, when the trail amount is still weak, the fraction of ants losing the trail is high and new food sources are likely to be discovered. The error level decreases as recruitment proceeds due to trail reinforcement by new recruits. The coupling of randomness in trail following behavior and competition between recruitment trails appears as a powerful strategy for optimizing foraging success.

An adaptive tuning of error has also been suggested as an interpretation of variations in the accuracy of honeybee dances (Towne and Gould, 1988). Information about food location differs in its level of accuracy as a function of the distance from the nest. At short distances (a few decameters), the dance is circular and contains no directional information. When food sources are remote, i.e. hundreds of meters from the hive, a bee recruiter is potentially able to code spatial information with a high level of accuracy through the direction of waggle runs. However, one curious feature of bee dancers is that their consecutive waggle runs may be produced with some directional scatter. A bee dancing for a less remote food source will perform relatively less precise waggle runs so that recruits will be distributed across several flower patches instead of being focused on where the communicating bee has been foraging (Weidenmuller and Seeley, 1999). The high-energy costs associated with flight may have favored a higher accuracy of communication about more remote food location. It is, however, suggested that imprecision in dance communication is caused primarily by physical constraints such as the ability of dancers to turn around quickly enough when the advertised site is nearby (Beekman *et al.*, 2005; see also Tanner and Visscher, 2006). The accuracy of directional information of food sources also changes according to the size of the resource. When the target of recruitment is a large flower patch, the dances show a lower directional accuracy and a higher angular divergence of consecutive waggle runs than in dances coding for a more definite spot in space such as a nestbox. This tuning of the dances' information content, which spreads recruits over a certain area, can be viewed as an adaptation to the typical spatial configuration of the recruitment targets (Weidenmuller and Seeley, 1999).

The modulation of the amount of trail pheromone laid by ants, as well as the fine-tuning of the angular divergences of bee dances, can both be viewed as striking adaptations for balancing the trade-off between an efficient exploitation of food and new food discoveries around the nest.

6 Spatial patterns

6.1 BEE LINES AND TRAIL NETWORKS

In both ants and honeybees, the spatial distribution of foraging routes follows that of food resources: routes leading to highly rewarding food patches will be preferentially foraged, whereas routes leading to poor or exhausted resources will be neglected. There are, however, several major differences in the spatial foraging patterns displayed by ant colonies or hives. As a whole, the pattern of foraging routes displayed by the hive appears as a juxtaposition of straight lines (beelines) between the nest and the food sources (Fig. 9B). These beelines effectively translate information encoded in the waggle dance. Even though some local deviations may occur due to the imprecision of the communication or to physical obstacles, newly recruited bees fly directly from the hive to the vicinity of a food source. Then, in the final stages of their flight, recruits proceed to search for the exact location of the flowers, their flight may become sinuous and their spatial information may be completed by odor and visual cues that help in the localization of exploited flowers (Tautz and Sandeman, 2003; Riley *et al.*, 2005; Arenas *et al.*, 2007). This searching behavior close to the flower patch accounts for the fact that the arrival of recruits at the source is often later than would be expected for a direct flight between the hive and the food source.

A major feature of honeybee lines is that the trajectory of each forager is individual. Before leaving the hive, each recruit has received, through the recruitment dance, the whole information needed to locate a flower patch. Each recruit exiting the hive then takes a flying route of its own. Ideally, this route should connect the hive entrance to the food patch in the straightest way possible in order to minimize the energy expenditure associated with flight. Hence, the spatial foraging pattern of the hive can be seen as a bundle of hundreds of bee lines that radiate out of the hive entrance and that are more concentrated in the angular zones leading to exploited flower patches. The spatial distribution of beelines is thus expected to be highly flexible: it should reflect changes in the direction coded in waggle runs and should closely follow changes in the spatial distribution of flower patches (Fig. 9B).

In ants, the spatial configuration of the foraging trails is quite different from that of beelines. Instead of being the mere juxtaposition of individual routes, trails often organize themselves into a network. Depending on the species, ants minimize the traveling time (Dussutour *et al.*, 2006) and/or the energy costs (Denny *et al.*, 2001) so that the final pattern of the trail network is deeply influenced by a set of spatial factors as well as by the history of the foraging activity (e.g. food discoveries, level of area marking, etc.) (Deneubourg *et al.*, 1990; Traniello *et al.*, 1991; Fourcassié and

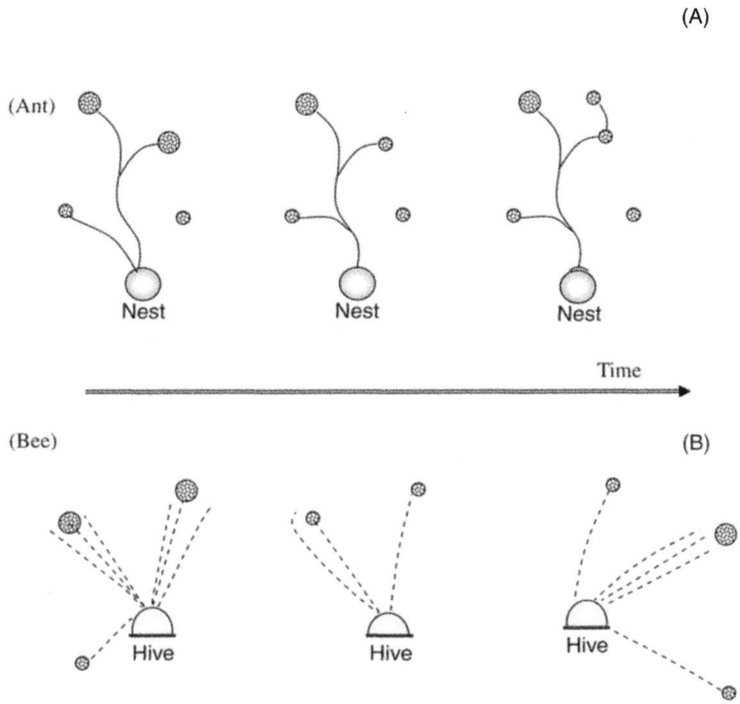

FIG. 9 Schematic representation of ant trails and beelines. Larger circles stand for larger food sources or larger flower patches. (A) Ant trails: Several resources may be concurrently exploited but some food sources remain neglected. Most of the foraging activity is focused on one food source. If depleted, this source becomes less foraged while another one starts attracting the foragers. A lateral trail that initially connects a food source to the nest entrance may, after a while, partially fuse with the main stream of foragers. A new trail can also originate from a previously exploited food source. (B) Beelines: The spatial distribution of beelines fits that of flower patches and radiate around the nest entrance. Flights become less frequent towards locations in which food patches become depleted. The direction of beelines is flexible and may markedly change when flowers blossom at quite different locations.

Deneubourg, 1994; Fourcassié *et al.*, 2003). For instance, if a new food source becomes available close to a source that is already foraged by ants, scouts will be more likely to discover it, they will lay a trail and, through path integration, they will follow a rather straight trajectory towards the nest. On their homeward journey, scouts may cross a well-established trail leading to another exploited food patch and may then decide to join the main stream of workers going back to the nest. Over time, successive visits of the foragers to the new food source will lead to the emergence of a new lateral trail branched onto the main foraging trail (Hölldobler and Wilson, 1970, 1990; Hölldobler and Möglich, 1980; Detrain *et al.*, 1991; Quinet and

Pasteels, 1991; Watmough and Edelstein-Keshet, 1995; Quinet *et al.*, 1997; Detrain *et al.*, 2000) (Fig. 9A). In the simple case of two close food sources, two straight independent trails leading to each food source scarcely coexist and most often merge into a common trail. As a result, the foraging activity becomes more concentrated on fewer paths as one travels from the food sites towards the nest entrance. The geometry of trail bifurcations within the network provides polarity information to foragers joining a trail, and these foragers can re-orientate themselves if they initially walk in the wrong direction (Jackson *et al.*, 2004). At the individual level, a tree-shaped trail network may lengthen the distance traveled by one forager, since its trajectory somewhat deviates from a straight line directly connecting the food source to the nest. However, at the society level, the building of a tree-shaped trail network greatly decreases the total trail length in comparison to a "star graphs" network with one direct trail going to each food source. Several benefits are associated with tree topologies: by reducing total network length, they facilitate the maintenance of foraging pathways (i.e. by actively cleaning or building construction around them; Howard, 2001), as well as their defense against competitors or predators.

The diversity of trail networks reported for ant species is a by-product of the interactions between trail-laying ants and any environmental factors that impede access to food sources or that may influence the probability of connecting two trails, such as the spatio-temporal availability of food resources (Deneubourg *et al.*, 1989; Franks *et al.*, 1991; Lopez *et al.*, 1994; Detrain *et al.*, 2000). For instance, two trails are more likely to merge into a common trunk trail when the two sources are persistent, close to each other, and/or remote from the nest. The higher sensitivity of the trail network to group-size effects is another major difference between ants and honeybees. In honeybees, an increase in the foraging effort will simply lead to an increase in the number of concurrent beelines. Conversely, in trail-laying ants, an increase in the number of workers may lead to qualitative changes in the spatial patterns of foraging routes. For instance, there is a shift from a diffuse foraging pattern to a tree-like trail network as the number of interacting foragers increases (Detrain *et al.*, 1991; Nicolis and Deneubourg, 1999). Such dependence on colony size is the typical signature of a self-organized process of pattern formation.

6.2 SELECTION OF THE SHORTEST PATH

For walking ants that encounter large obstacles (stones, crevices, etc.) compared to their tiny size, recruitment trails not only prevent them from losing their way but also allow them to select the shortest route towards the feeding site. While the choice of one path is random over a bridge with two equal branches, most of the foragers follow the shortest path of a bridge with branches of unequal length (Goss *et al.*, 1989; Vittori *et al.*, 2006).

This adaptive selection of the shortest path occurs even though individual ants have no knowledge of relative path lengths, as well as no memory of the routes they have previously taken. For ants that lay a trail in both directions (from the nest to the food source and vice versa), this selection is simply due to the shorter trip duration and hence to the increased rate of trail reinforcement over the short path. In some ant species that lay a trail only during their homeward trip (Beckers *et al.*, 1992b), the selection of the shortest path is also due to the U-turns made by workers that double back on a trail, return to the common branch point and then follow the second branch. Because when an ant is backtracking after its U-turn it does not lay down trail pheromone, the higher frequency of U-turns along the long path ultimately leads to a lower amount of pheromone over the longest branch. The adaptive selection of the shortest path thus arises "automatically" at the colony level from differences in the recruitment dynamics depending on branch length. This results from the synergistic effect of a shorter duration of foraging trips and of a lower frequency of U-turns on the rate of trail reinforcement.

It is difficult to find the counterpart for the honeybees. Firstly, the spatial information provided by the orientation of waggle dances is, by essence, the shortest flying route connecting the hive to the flower patch, since it results from a quite efficient path integration of the foraging journey by the recruiter (Collett and Collett, 2004, 2006; Collett *et al.*, 2006). Nevertheless, in the presence of obstacles, short cuts and shorter-lasting foraging journeys may be achieved on an individual basis by honeybee foragers that are more experienced about how to reach the flower patch.

7 Conclusions

7.1 WHAT ABOUT OTHER GROUP-LIVING ARTHROPODS?

Collective decision-making can be as accurate and effective as decisions made by solitary animals, even by those famous for their well-developed capacities for information processing, such as vertebrates (Krebs and Davies, 1984). The strength of collective decision-making is that adaptive choices may arise automatically without the need for any leaders, which would have a synoptic view of all environmental opportunities. Consensus decisions can be reached even when the individuals do not know how the information they possess compares with that of conspecifics. Many examples of collective decision-making have been reported in the literature not only for ants or honeybees (Hölldobler and Wilson, 1990; Seeley, 1995; Theraulaz and Spitz, 1997; Bonabeau *et al.*, 1997; Camazine *et al.*, 2001), but also for other group-living organisms including bacteria, which may achieve quorum sensing and behave collectively (Miller and Bassler, 2001;

Ben Jacob *et al.*, 2004). Likewise, complex patterns have been reported in fungal networks (Bebber *et al.*, 2007) and slime-moulds (Shirakawa and Gunji, 2007). In this respect, group-living arthropods display amazing collective patterns and coordinate their efforts efficiently even though their social organization is less complex than that of social insects (Fitzgerald, 1995; Fitzgerald and Costa, 1999; Deneubourg *et al.*, 2002; Costa, 2006). The choice of a common shelter or the cooperative exploitation of one resource by several individuals is then guided not only by cues related to the quality of these resources, but also by recruiting and/or arresting signals emitted by conspecifics. Hundred of examples of aggregative behavior, which rely on pheromones, have been reported for species living in very different ecological situations [e.g. *Triatoma* kissing bugs (Lorenzo and Lazzari, 1996); cockroaches (Miller and Koehler, 2000; Said *et al.*, 2005); *Drosophila* fruit flies (Takahashi, 2006); lobsters (for a review see Wertheim *et al.*, 2005; Horner *et al.*, 2006)]. The active recruitment of conspecifics toward a common place can also be mechanically mediated, as has been shown in gregarious larvae of herbivorous insects, such as treehoppers, which use vibrations to orient themselves and eventually to move from low- to high-quality sites (Cocroft, 2005).

Although collective patterns of aggregation are widely reported in arthropods, there are surprisingly few experiments investigating how collective choices emerge among group members and how they are influenced by the ways conspecifics interact with each others. The use of silk impregnated with pheromone is one of the most studied long-range interactions in gregarious arthropods. Social caterpillars (Fitzgerald, 1995) and mites (Yano, 2008) may lay silk threads that attract conspecifics in a manner similar to ant trails. The tuning of the amount of pheromone released over the silk enables caterpillars to collectively select richer leaf patches as well as to reach a consensus about a common resting site at the end of a day's activity (Fitzgerald and Peterson, 1983; Fitzgerald, 1995; Ruf *et al.*, 2001). Moreover, like ants or stingless bees, forest tent caterpillars, *Malacosoma disstria*, show a lack of flexibility in their collective decision-making. Indeed, when faced with a nutritionally balanced and a less profitable, unbalanced source placed at opposite ends of a bridge, the colony of caterpillars as a whole, randomly discovers one of the two resources, goes on following the recruitment silk threads and becomes trapped for at least one day in the exploitation of the resource discovered first independently of its quality (Dussutour *et al.*, 2007a,b).

Group-living spiders provide another striking example of collective decision-making. The strong correlation between the amount of exploratory webs and the localization of the spiders' population suggests that conspecifics' draglines are used as a recruitment means for colony formation. Actually, groups of *Anelosimus studiosus* spiders can only follow individuals sequentially released at a common spot when draglines are left in place on the substrate (Furey, 1998). The silk network orients the movement of individual spiders: it acts as a recruiting structure and, due to the pattern of dragline attachments, it may lead in a binary choice setup to the collective selection of only one aggregation site (Saffre *et al.*, 1999; Jeanson *et al.*, 2004b).

The emergence of collective patterns does not necessarily imply the emission of recruiting signals and may also be triggered by short-range physical contact. For instance, the aggregation of cockroaches under a common shelter is due to their perception of contacts with conspecifics, in which the individual probability of moving decreases with the number of neighbors resting under a shelter (Ame *et al.*, 2004, 2006; Jeanson *et al.*, 2005; Halloy *et al.*, 2007). Other spatial patterns (other than binary choices) may be seen as by-products of collective choices in which the group members choose between two or more mutually exclusive actions. For instance, in groups of locusts of increasing density, a transition occurs from a disordered movement of individuals within the group to a highly aligned collective motion. This transition may be explained by mechanisms and decision rules similar to those described previously (Buhl *et al.*, 2006).

7.2 NON-LINEARITY OF INFORMATION FLOW: THE KEY-STONE OF COLLECTIVE DECISION-MAKING

The ways individuals interact with each other and give rise to collective responses vary, depending on the species concerned and may involve chemical signals, physical contacts or both types of stimuli (Camazine and Sneyd, 1991; Seeley *et al.*, 1991; Visscher and Camazine, 1999a,b; Visscher, 2007). Insect societies such as the hive or the ant nest are real life models that offer a unique opportunity to investigate (1) how collective structures emerge in complex biological systems made up of numerous individuals, and (2) how collective behavior may be shaped by the properties of one species' form of communication. Due to the specificities of chemical trails versus dances, ant nests and hives differ in the amplifying potential of their recruiting process. Likewise, in other group-living arthropods, a difference in the level of non-linearity of one individual's response to recruiting stimuli may have major consequences on the collective patterns displayed by the whole group. Usually, chemical signals are characterized by a pheromone decay rate and a receiver response curve that are both highly non-linear. Such a non-linearity of information flow – as found in ants' chemical trails – favors the focusing of the group at one or a few sites, but restrains the flexibility of the group to shift to more profitable resources. Hence, a non-linearity of receivers' response is expected to be found in species that benefit from cooperation in the exploitation or in the defense of resources, as well as in species exploiting resources that are rather stable in time and space. Conversely, tactile signals such as the honeybee dances are expected to show less non-linear properties than chemical ones since the rates of contacts should be proportional to the number of interacting individuals. The individual response to contact rates may nevertheless be non-linear as found in the case of nest-moving recruitment (Pratt *et al.*, 2002; Seeley and Visscher, 2003; Jeanson *et al.*, 2004a; Pratt and Sumpter, 2006) or in the case of aggregation patterns in ants (Gordon *et al.*, 1993; Depickere *et al.*, 2004; Nicolis *et al.*, 2005) and cockroaches (Ame *et al.*, 2004, 2006; Jeanson *et al.*, 2005). Whatever the type of between-nestmate interactions, a low level of non-linearity favors an even distribution of individuals among identical resources. It also allows a closer fitting between the number of animals allocated to one resource and its relative profitability (as predicted by an ideal free distribution). Finally, it facilitates higher flexibility in-group decisions when environmental resources are changing. This should benefit species that (1) exploit resources, the availability of which is highly variable in time and space, (2) that do not need to monopolize resources against competitors, and (3) that do not need to cooperatively defend food sites because individuals are well protected against predators (e.g. by defensive secretions or effective weapons, such as in honeybees).

The evolution of communication in gregarious insects has been presumably steered by whether, for a particular colony sizes and habitat, a recruitment increases food collection, favors the monopolization of resources or prevents attacks by predators. Then, the payoff of one type of recruitment should vary depending on the foraging ecology of the species. In particular, recruitment mechanisms with a higher level of non-linearity seem well-adapted to species that benefit from cooperation and collective selection of a subset of resources at the expense of a lower flexibility to changing environmental conditions. In this respect, being able to shift between signals that differ in their level of non-linearity may be advantageous for one species. For example, within the same recruitment process, some ant species may change the way they communicate with each other: scouts that have discovered a new source first perform group recruitment to activate nestmates and thereafter shift towards mass recruitment by chemical trails. Group recruitment (such as tandem recruitment or honeybee dances) is characterized by a low level of non-linearity: the number of recruited foragers is largely proportional to the number of successful scouts, as the group of recruits has to be physically guided by a recruiting leader until they reach the food source (e.g. *Tetramorium* ant species, Verhaeghe, 1982; see also Hölldobler and Wilson, 1990; de Biseau *et al.*, 1994). Over the course of food exploitation, mass recruitment and non-linear properties of chemical trail recruitment become prevalent, with most of the foragers being concentrated on a few food sources (Pasteels *et al.*, 1987). This double recruitment process confers a higher flexibility to the ant nest. Any patrolling ant which has missed the first food source and that discovers a more profitable one, can behave as a leader: it guides groups of recruits towards this new resource, counteracts the attractiveness of the chemical trail leading to the first exploited target and, as observed in honeybees, facilitates the shift of the foraging activity towards richer food sources (Beckers *et al.*, 1990).

A higher complexity of communication is an additional means for a group-living species to improve flexibility, as well as diversity of collective patterns. In the case of foraging, recruitment signals are often a blend of molecules with different life-times, operating at different stages of food exploitation (see e.g. de Biseau *et al.*, 1991; Detrain and Cammaerts, 1991). Another example is the synergistic effect of recruitment trail and home-range marking, the latter being ant density-dependent (Devigne and Detrain, 2002; Devigne *et al.*, 2004; Devigne and Detrain, 2006): this means that workers are recruited to areas as a function of their frequentation level and hence their potential interest as foraging sites. Foraging efficiency can also be improved if repellent pheromones are used to mark unrewarding areas (Giurfa, 1993; Stickland *et al.*, 1999; Robinson *et al.*, 2005). A challenge for future research will be to investigate to what extent the level of non-linearity of one species' interactions matches its need for cooperative behavior or for flexibility to adapt to environmental changes.

7.3 WHAT ABOUT INDIVIDUALITIES AND THEIR IMPACT ON COLLECTIVE DECISION-MAKING?

Whatever their levels of sociality, group-living species share a common property: the potential for amplifying, at the group level, individual traits or individual preferences. For instance, when ants are given the choice on a diamond-shaped bridge between a branch with or without a wall, about 65% of the foragers will prefer to follow the walled branch in the absence of a recruitment trail. But, as recruitment proceeds, this propensity of ants will increase and will lead up to 85% of the foragers to choose the walled path (Dussutour *et al.*, 2005a). This demonstrates that non-linearity in between-nestmate interactions may enhance even slight preferences of the individuals and lead to much more clear-cut choices by the group as a whole. Likewise, a synergy may exist between individual memory or fidelity to an area or to a given resource and the amplifying potential of recruitment; so that the specialization of a small number of recruiters will focus most of the colony in this particular sector or onto this food source. Amplifying phenomena that underlie most collective decisions are thus powerful means by which slight changes in physiological preferences (e.g. thermopreferendum values), changes in behavioral traits (e.g. thigmotaxis) or changes in individual knowledge of the environment can have deep consequences on the way a species will live as a group, shelter exclusively in a peculiar resting site, travel over a determined foraging path, and focus its foraging effort on a given resource.

One may erroneously think about collective decision-making and idiosyncrasy as being opposite concepts (Deneubourg *et al.*, 1999). In fact, individual memory and individual experience do not necessarily prevent the reaching of a consensus, but instead make the patterns that will emerge at the colony level more complex and diverse (Couzin *et al.*, 2005). When "conflicts" exist between the preferences of different individuals, there are multiple possible outcomes. One is the splitting of the group with individuals possibly losing the benefits associated with being part of a large group (Krause and Ruxton, 2002) but with the group enlarging its occupation range of the environment. Second is the reaching of a consensus with a small proportion of informed individuals acting as leaders and guiding a group primarily composed of naïve individuals towards a target location. Research should now focus on how individualities may be enhanced at the group level by positive feedbacks and give rise to a multiplicity of adaptive collective patterns. As a corollary, one should question how species that benefit from maintaining group cohesion have evolved mechanisms resolving potential conflicts between different individual preferences. Part of the answer could be found in the ratio of recruiters (or informed individuals) versus naive ones, as well as in the level of non-linearity of their communication. Answers to these questions will ultimately provide new insights into the evolution of communication and cooperation in group-living animals.

Acknowledgments

This work has been financially supported by the "Sillages" project (ANR-05-BLAN-017701) and the "ECagents" project (Future and Emerging Technologies program of the European Community, IST-1940). The authors would like to thank F. Detournay and L. Dekelver for their editing assistance, as well as the two anonymous referees for their constructive comments. CD and JLD are Senior Research Associates from the Belgian National Fund for Scientific Research.

References

Afik, O., Dag, A. and Shafir, S. (2008). Honeybee, *Apis mellifera*, round dance is influenced by trace components of floral nectar. *Anim. Behav.* **75**, 371–377.

Ame, J. M., Rivault, C. and Deneubourg, J. L. (2004). Cockroach aggregation based on strain odour recognition. *Anim. Behav.* **68**, 793–801.

Ame, J. M., Halloy, J., Rivault, C., Detrain, C. and Deneubourg, J. L. (2006). Collegial decision making based on social amplification leads to optimal group formation. *Proc. Natl. Acad. Sci. USA* **103**, 5835–5840.

Anderson, C., Boomsma, J. J. and Bartholdi, J. J. (2002). Task partitioning in insect societies: bucket brigades. *Insect. Soc.* **49**, 171–180.

Arenas, A., Fernandez, V. M. and Farina, W. M. (2007). Floral odor learning within the hive affects honeybees' foraging decision. *Naturwissenschaften* **94**, 218–222.

Bebber, D. P., Darrah, P. R., Hynes, J., Boddy, L. and Fricker, M. D. (2007). Biological solutions to transport network design. *Proc. R. Soc. B.* **274**, 2307–2315.

Beckers, R., Goss, S., Deneubourg, J. L. and Pasteels, J. M. (1989). Colony size, communication and ant foraging strategies. *Psyche* **96**, 239–256.

Beckers, R., Deneubourg, J. L., Goss, S. and Pasteels, J. M. (1990). Collective decision making through food recruitment. *Insect. Soc.*, 258–267.

Beckers, R., Deneubourg, J. L. and Goss, S. (1992a). Trail laying behavior during food recruitment in the ant *Lasius niger*. *J. Theor. Biol.* **59**, 397–415.

Beckers, R., Deneubourg, J. L. and Goss, S. (1992b). Trails and U-turns in the selection of a path by the ant *Lasius niger*. *Insect. Soc.* **59**, 397–415.

Beckers, R., Deneubourg, J. L. and Goss, S. (1993). Modulation of trail-laying in the ant *Lasius niger* and its role in the collective selection of a food source. *J. Insect. Behav.* **6**, 751–759.

Beekman, M., Sumpter, D. J. T., Seraphides, N. and Ratnieks, F. L. W. (2004). Comparing foraging behaviour of small and large honey-bee colonies by decoding waggle dances made by foragers. *Funct. Ecol.* **18**, 829–835.

Beekman, M., Doyen, L. and Oldroyd, B. P. (2005). Increase in dance imprecision with decreasing foraging distance in the honey bee *Apis mellifera* L. is partly explained by physical constraints. *J. Comp. Physiol. A Neuroethol. Sens. Neural Behav. Physiol.* **191**, 1107–1113.

Ben Jacob, E., Becker, I., Shapiro, Y. and Levine, H. (2004). Bacterial linguistic communication and social intelligence. *Trends Microbiol.* **12**, 366–372.

Beshers, S. N. and Fewell, J. H. (2001). Models of division of labor in social insects. *Ann. Rev. Entomol.* **46**, 413–414.

Biesmeijer, J. C. and Ermers, M. C. W. (1999). Social foraging in stingless bees: how colonies of *Melipona fasciata* choose among nectar sources. *Behav. Ecol. Sociobiol.* **46**, 129–140.

Boinski, S. and Garber, P. A. (2000). *On the Move. How and Why Animals Travel in Group*. Chicago: University of Chicago Press.

Bonabeau, E., Theraulaz, G., Deneubourg, J. L., Aron, S. and Camazine, S. (1997). Self-organization in social insects. *Trends Ecol. Evol.* **12**, 188–193.

Buhl, J., Sumpter, D. J. T., Couzin, I. D., Hale, J. J., Despland, E., Miller, E. R. and Simpson, S. J. (2006). From disorder to order in marching locusts. *Science* **312**, 1402–1406.

Calenbuhr, V., Chretien, L., Deneubourg, J.-L. and Detrain, C. (1992). A model for osmotropotactic orientation (II). *J. Theor. Biol.* **158**, 395–407.

Calenbuhr, V. and Deneubourg, J.-L. (1992). A model for osmotropotactic orientation (I). *J. Theor. Biol.* **158**, 359–393.

Camazine, S. and Sneyd, J. (1991). A model of collective nectar source selection by honey-bees: self-organization through simple rules. *J. Theor. Biol.* **149**, 547–571.

Camazine, S., Crailsheim, K., Hrassnigg, N., Robinson, G. E., Leonhard, B. and Kropiunigg, H. (1998). Protein trophallaxis and the regulation of pollen foraging by honey bees (*Apis mellifera* L.). *Apidologie* **29**, 113–126.

Camazine, S., Deneubourg, J. L., Franks, N. R., Sneyd, J., Theraulaz, G. and Bonabeau, E. (2001). *Self-Organization in Biological Systems*. Princeton: Princeton University Press.

Cassill, D. L. (2003). Rules of supply and demand regulate recruitment to food in an ant society. *Behav. Ecol. Sociobiol.* **54**, 441–450.

Cassill, D. L. and Tschinkel, W. R. (1999a). Regulation of diet in the fire ant, *Solenopsis invicta*. *J. Insect. Behav.* **12**, 307–328.

Cassill, D. L. and Tschinkel, W. R. (1999b). Information flow during social feeding in ant societies. In: *Information Processing in Social Insects* (eds Detrain, C., Deneubourg, J. L. and Pasteels, J. M.), pp. 69–81. Basel: Birkhaüser Verlag.

Cocroft, R. B. (2005). Vibrational communication facilitates cooperative foraging in a phloem-feeding insect. *Proc. R. Soc. B Biol. Sci.* **272**, 1023–1029.

Collett, T. S. and Collett, M. (2004). How do insects represent familiar terrain? Paris. *J. Physiol. (Paris)* **98**, 259–264.

Collett, M. and Collett, T. S. (2006). Insect navigation: no map at the end of the trail? *Curr. Biol.* **16**, R48–R51.

Collett, M., Collett, T. S. and Srinivasan, M. V. (2006). Insect navigation: measuring travel distance across ground and through air. *Curr. Biol.* **16**, R887–R890.

Condorcet, M. (1785). *Essai sur l'application de l'analyse à la probabilité des décisions rendues à la pluralité des voix*. Paris: Imprimerie Royale.

Conradt, L. and Roper, J. T. (2005). Consensus decision making in animals. *Trends Ecol. Evol.* **20**, 449–456.

Costa, J. T. (2006). *The Other Insect Societies*. Cambridge, MA: Belknap Press of Harvard University Press.

Couzin, I. D., Krause, J., Franks, N. R. and Levin, S. A. (2005). Effective leadership and decision-making in animal groups on the move. *Nature* **433**, 513–516.

Czechowski, W. (1984). Tournaments and raids in *Lasius niger* (L.) (Hymenoptera, Formicidae). *Ann. Zool.* **38**, 81–90.

Danchin, E., Giraldeau, L. A., Valone, T. J. and Wagner, R. H. (2004). Public information: from nosy neighbors to cultural evolution. *Science* **305**, 487–491.

de Biseau, J. C. and Pasteels, J. M. (1994). Regulated food recruitment through individual behavior of scouts in the ant, *Myrmica sabuleti* (Hymenoptera: Formicidae). *J. Insect. Behav.* **7**, 767–777.

de Biseau, J. L., Deneubourg, J. L. and Pasteels, J. M. (1991). Collective flexibility during mass recruitment in the ant *Myrmica sabuleti* Hymenoptera Formicidae. *Psyche* **98**, 323–336.

de Biseau, J. C., Schuiten, M., Pasteels, J. M. and Deneubourg, J. L. (1994). Respective contributions of leader and trail during recruitment to food in *Tetramorium bicarinatum* (Hymenoptera, Formicidae). *Insect. Soc.* **41**, 241–254.

De Marco, R. J. and Farina, W. M. (2001). Changes in food profitability affect the trophallactic and dance behaviour of forager honeybees (*Apis mellifera*). *Behav. Ecol. Sociobiol.* **50**, 441–449.

Deneubourg, J. L. and Goss, S. (1989). Collective patterns and decision-making. *Ethol. Ecol. Evol.* **1**, 295–311.

Deneubourg, J. L., Pasteels, J. M. and Verhaeghe, J. C. (1983). Probabilistic behaviour in ants: a strategy of errors? *J. Theor. Biol.* **105**, 259–271.

Deneubourg, J. L., Goss, S., Pasteels, J. M., Fresneau, D. and Lachaud, J. P. (1987). Self-organization mechanisms in ant societies (II): learning in foraging and division of labor. *Experientia Suppl.* **54**, 177–196.

Deneubourg, J. L., Goss, S., Franks, N. and Pasteels, J. M. (1989). The blind leading the blind: modeling chemically mediated army ant raid patterns. *J. Insect. Behav.* **2**, 719–725.

Deneubourg, J. L., Aron, S., Goss, S. and Pastels, J. M. (1990). The self-organizing exploratory pattern of the Argentine ant. *J. Insect. Behav.* **3**, 159–168.

Deneubourg, J. L., Camazine, S. and Detrain, C. (1999). Self-organization or individual complexity: a false dilemma or a true complementarity? In: *Information Processing in Social Insects* (eds Detrain, C., Deneubourg, J. L. and Pasteels, J. M.), pp. 401–408. Basel: Birkhaüser.

Deneubourg, J.-L., Lioni, A. and Detrain, C. (2002). Dynamics of aggregation and emergence of cooperation. *Biol. Bull.* **202**, 262–267.

Denny, A. J., Wright, J. and Grief, B. (2001). Foraging efficiency in the wood ant, *Formica rufa*: is time of the essence in trail following? *Anim. Behav.* **62**, 139–146.

Depickere, S., Fresneau, D. and Deneubourg, J. L. (2004). A basis for spatial and social patterns in ant species: dynamics and mechanisms of aggregation. *J. Insect. Behav.* **17**, 81–97.

Detrain, C. and Cammaerts, M. C. (1991). A new pheromone in the ant *Pheidole pallidula*. *Behav. Processes* **24**, 123–132.

Detrain, C. and Deneubourg, J. L. (1997). Scavenging by *Pheidole pallidula*: a key for understanding decision-making systems in ants. *Anim. Behav.* **53**, 537–547.

Detrain, C. and Deneubourg, J. L. (2002). Complexity of environment and parsimony of decision rules in insect societies. *Biol. Bull.* **202**, 268–274.

Detrain, C. and Deneubourg, J. L. (2006). Self-organized structures in a superorganism: do ants "behave" like molecules? *Phys. Life Rev.* **3**, 162–187.

Detrain, C. and Pasteels, J. M. (1991). Caste differences in behavioral thresholds as a basis for polyethism during food recruitment in the ant *Pheidole pallidula*. *J. Insect. Behav.* **4** (2), 157–177.

Detrain, C. and Pasteels, J. M. (1992). Caste polyethism and collective defense in the ant *Pheidole pallidula*: the outcome of quantitative differences in recruitment. *Behav. Ecol. Sociobiol.* **29**, 405–412.

Detrain, C., Pasteels, J. M. and Deneubourg, J. L. (1988). Polyéthisme dans le tracé et le suivi de la piste chez *Pheidole pallidula*. *Actes des Colloques Insectes Sociaux* **4**, 87–94.

Detrain, C., Deneubourg, J. L., Goss, S. and Quinet, Y. (1991). Dynamics of collective exploration in the ant *Pheidole pallidula*. *Psyche* **98**, 21–32.

Detrain, C., Deneubourg, J. L. and Pasteels, J. M. (1999). Decision-making in foraging by social insects. In: *Information Processing in Social Insects* (eds Detrain, C., Deneubourg, J. L. and Pasteels, J. M.), pp. 331–354. Basel: Birkhaüser Verlag.

Detrain, C., Tasse, O., Versaen, M. and Pasteels, J. M. (2000). A field assessment of optimal foraging in ants: trail patterns and seed retrieval by the European harvester ant *Messor barbarus*. *Insect. Soc.* **47**, 56–62.

Detrain, C., Natan, C. and Deneubourg, J. L. (2001). The influence of the physical environment on the self-organised foraging patterns of ants. *Naturwissenschaften* **88**, 171–174.

D'Ettore, P. (2007). Evolution of sociality: you are what you learn. *Curr. Biol.* **17**, 766–768.

Devigne, C. and Detrain, C. (2002). Collective exploration and area marking in the ant *Lasius niger*. *Insect. Soc.* **49**, 357–362.

Devigne, C. and Detrain, C. (2006). How does food distance influence foraging in the ant *Lasius niger*: the importance of home-range marking. *Insect. Soc.* **53**, 46–55.

Devigne, C., Renon, A. J. and Detrain, C. (2004). Out of sight but not out of mind: modulation of recruitment according to home range marking in ants. *Anim. Behav.* **67**, 1023–1029.

De Vries, H. and Biesmeijer, J. C. (1998). Modelling collective foraging by means of individual behaviour rules in honey-bees. *Behav. Ecol. Sociobiol.* **44**, 109–124.

De Vries, H. and Biesmeijer, J. C. (2002). Self-organization in collective honeybee foraging: emergence of symmetry breaking, cross inhibition and equal harvest-rate distribution. *Behav. Ecol. Sociobiol.* **51**, 557–569.

Diaz, P. C., Gruter, C. and Farina, W. M. (2007). Floral scents affect the distribution of hive bees around dancers. *Behav. Ecol. Sociobiol.* **61**, 1589–1597.

Dussutour, A., Fourcassie, V., Helbing, D. and Deneubourg, J. L. (2004). Optimal traffic organization in ants under crowded conditions. *Nature* **428**, 70–73.

Dussutour, A., Deneubourg, J. L. and Fourcassie, V. (2005a). Amplification of individual preferences in a social context: the case of wall-following in ants. *Proc. R. Soc. B Biol. Sci.* **272**, 705–714.

Dussutour, A., Deneubourg, J. L. and Fourcassie, V. (2005b). Temporal organization of bi-directional traffic in the ant *Lasius niger* (L.). *J. Exp. Biol.* **208**, 2903–2912.

Dussutour, A., Nicolis, S. C., Deneubourg, J. L. and Fourcassié, V. (2006). Collective decisions in ants when foraging under crowded conditions. *Behav. Ecol. Sociobiol.* **61**, 17–30.

Dussutour, A., Beshers, S., Deneubourg, J.-L. and Fourcassie, V. (2007a). Crowding increases foraging efficiency in the leaf-cutting ant *Atta colombica. Insect. Soc.* **54**, 158–165.

Dussutour, A., Simpson, S. J., Despland, E. and Colasurdo, N. (2007b). When the group denies individual nutritional wisdom. *Anim. Behav.* **74**, 931–939.

Dyer, F. C. (2002). The biology of the dance language. *Annu. Rev. Entomol.* **47**, 917–949.

Dyer, F. C. and Seeley, T. D. (1991). Dance dialects and foraging range in three Asian honey-bee species. *Behav. Ecol. Sociobiol.* **28**, 227–233.

Dyer, J. R. G., Ioannou, C. C., Morrell, L., Crot, D. P., Couzin, I. D., Waters, D. A. and Krause, J. (2008). Consensus decision making in human crowds. *Anim. Behav.* **75**, 461–470.

Edelstein-Keshet, L., Watmough, J. and Ermentrout, G. B. (1995). Trail following in ants: individual properties determine population behavior. *Behav. Ecol. Sociobiol.* **36**, 119–133.

Farina, W. M., Gruter, C. and Diaz, P. C. (2005). Social learning of floral odours inside the honeybee hive. *Proc. R. Soc. B Biol. Sci.* **272**, 1923–1928.

Fitzgerald, T. D. (1995). *The Tent Caterpillars*. Ithaca: Cornell University Press.

Fitzgerald, T. D. and Costa, J. T. (1999). Collective behavior in social caterpillars. In: *Information Processing in Social Insects* (eds Detrain, C., Deneubourg, J. L. and Pasteels, J. M.), pp. 379–400. Basel: Birkhaüser Verlag.

Fitzgerald, T. D. and Peterson, S. C. (1983). Elective recruitment by the Eastern tent caterpillar (*Malacosoma americanum*). *Anim. Behav.* **31**, 417–423.

Fourcassié, V. and Deneubourg, J. L. (1994). The dynamics of collective exploration and trail-formation in *Monomorium pharaonis*: experiments and model. *Physiol. Entomol.* **19**, 291–300.

Fourcassié, V., Bredard, C., Volpatti, K. and Theraulaz, G. (2003). Dispersion movements in ants: spatial structuring and density dependent effects. *Behav. Processes* **63**, 33–43.

Franks, N. R. and Fletcher, C. R. (1983). Spatial patterns in army ant foraging and migration: *Eciton burchelli* on Barro Colorado Island, Panama. *Behav. Ecol. Sociobiol.* **12**, 261–270.

Franks, N. R. and Partridge, L. W. (1993). Lanchester battles and the evolution of combat in ants. *Anim. Behav.* **45**, 197–199.

Franks, N. R. and Richardson, T. (2006). Teaching in tandem-running ants. *Nature* **439**, p. 153.

Franks, N. R., Gomez, N., Goss, S. and Deneubourg, J. L. (1991). The blind leading the blind in army ant raid patterns: testing the model of self organisation (Hymenoptera: Formicidae). *J. Insect. Behav.* **4**, 583–607.

Franks, N. R., Pratt, S. C., Mallon, E. B., Britton, N. F. and Sumpter, D. J. T. (2002). Information flow, opinion polling and collective intelligence in house-hunting social insects. *Philos. Trans. R. Soc. B Biol. Sci.* **357**, 1567–1583.

Furey, F. E. (1998). Two cooperatively social populations of the theridiid spider *Anelosimus studiosus* in a temperate region. *Anim. Behav.* **55**, 727–735.

Gaertner, W. (2001). *Domain Conditions in Social Choice Theory*. Cambridge: Cambridge University Press.

Gil, M. and Farina, W. M. (2002). Foraging reactivation in the honeybee *Apis mellifera* L.: factors affecting the return to known nectar sources. *Naturwissenschaften* **89**, 322–325.

Giraldeau, L. A. and Caraco, T. (2000). *Social Foraging Theory*. Princeton, NJ: Princeton University Press.

Giurfa, M. (1993). The repellent scent-mark of the honeybee *Apis mellifera-Ligustica* and its role as communication cue during foraging. *Insect. Soc.* **40**, 59–67.

Gordon, D. M., Paul, R. E. and Thorpe, K. (1993). What is the function of encounter patterns in ant colonies? *Anim. Behav.* **45**, 1083–1100.

Goss, S. and Deneubourg, J. L. (1989). The self-organising clock pattern of *Messor pergandei* Formidae Myrmicinae. *Insect. Soc.* **36**, 339–347.

Goss, S., Aron, S., Deneubourg, J. L. and Pasteels, J. M. (1989). Self-organized shortcuts in the argentine ant. *Naturwissenschaften* **76**, 579–581.

Gould, J. L. (1976). The dance-language controversy. *Q. Rev. Biol.* **51**, 211–244.

Grassé, P. P. (1959). La reconstruction du nid et les coordinations interindividuelles chez *Bellicositermes natalensis* et *Cubitermes sp.*, La théorie de la stigmergie: essais d'interprétation du comportement des termites constructeurs. *Insect. Soc.* **6**, 41–84.

Guzmannovoa, E. and Page, R. E. (1994). Genetic dominance and worker interactions affect honeybee colony defense. *Behav. Ecol.* **5**, 91–97.

Halloy, J., Sempo, G., Caprari, G., Rivault, C., Asadpour, M., Tâche, F., Saïd, I., Durier, V., Canonge, S., Amé, J.-M., Detrain, C., Correll, N., Martinoli, A., Mondada, F., Siegwart, R. and Deneubourg, J.-L. (2007). Social integration of robots into groups of cockroaches to control self-organized choices. *Science* **318**, 1155–1158.

Hangartner, W. (1967). Spezifität und Inaktivierung des Spurpheromons von *Lasius fuliginosus* Latr. und Orientierung der Arbeiterinnen im Duftfeld. *J. Comp. Physiol. A Neuroethol. Sens. Neural Behav. Physiol.* **57**, 103–136.

Hangartner, W. (1969). Structure and variability of the individual odor trail in *Solenopsis geminata* (Hymenoptera formicidae). *J. Comp. Physiol. A Neuroethol. Sens. Neural Behav. Physiol.* **62**, 111–120.

Heidecker, J. L. and Leuthold, R. H. (1984). The organization of collective foraging in the haverster termite *Hodotermes mossambicus* (Isoptera). *Behav. Ecol. Sociobiol.* **14**, 195–202.

Hölldobler, B. and Möglich, M. (1980). The foraging system of *Pheidole militicida* (Hymenoptera, Formicidae). *Insect. Soc.* **27**, 237–264.

Hölldobler, B. and Wilson, E. O. (1970). Recruitment trails in the harvester ants *Pogonomyrmex badius*. *Psyche* **77**, 385–399.

Hölldobler, B. and Wilson, E. O. (1990). *The Ants*. Berlin: Springer Verlag.

Horner, A. J., Nickles, S. P., Weissburg, M. J. and Derby, C. D. (2006). Source and specificity of chemical cues mediating shelter preference of Caribbean spiny lobsters (*Panulirus argus*). *Biol. Bull.* **211**, 128–139.

Howard, J. J. (2001). Costs of trail construction and maintenance in the leaf-cutting ant *Atta columbica*. *Behav. Ecol. Sociobiol.* **49**, 348–356.

Howard, D. F. and Tschinkel, W. R. (1981). Internal distribution of liquid foods in isolated workers of the fire ant, *Solenopsis invicta*. *J. Insect. Physiol.* **27**, 67–74.

Howard, J. J., Henneman, M. L., Cronin, G., Fox, J. A. and Hormiga, G. (1996). Conditioning of scouts and recruits during foraging by a leafcutting ant, *Atta colombica*. *Anim. Behav.* **52**, 299–306.

Hunt, G. J., Guzman-Novoa, E., Uribe-Rubio, J. L. and Prieto-Merlos, D. (2003). Genotype-environment interactions in honeybee guarding behaviour. *Anim. Behav.* **66**, 459–467.

Jackson, D. E., Holcombe, M. and Ratnieks, F. L. W. (2004). Trail geometry gives polarity to ant foraging networks. *Nature* **432**, 907–909.

Jarau, S., Hrncir, M., Zucchi, R. and Barth, F. G. (2003). Effectiveness of recruitment behaviour in stingless bees (Apidae, Meliponini). *Insect. Soc.* **50**, 365–374.

Jarau, S., Hrncir, M., Zucchi, R. and Barth, F. G. (2004). A singles bee uses labial gland secretions for scent trail communication (*Trigona recursa* Smith 1863). *J. Comp. Physiol. A* **190**, 233–239.

Jeanson, R., Ratnieks, F. L. and Deneubourg, J. L. (2003). Pheromone trail decay rates on different substrates in the Pharaoh ants, *Monomorium pharaonis*. *Physiol. Entomol.* **28**, 192–198.

Jeanson, R., Deneubourg, J. L., Grimal, A. and Theraulaz, G. (2004a). Modulation of individual behavior and collective decision-making during aggregation site selection by the ant *Messor barbarus*. *Behav. Ecol. Sociobiol.* **55**, 388–394.

Jeanson, R., Deneubourg, J. L. and Theraulaz, G. (2004b). Discrete dragline attachment induces aggregation in spiderlings of a solitary species. *Anim. Behav.* **67**, 531–537.

Jeanson, R., Rivault, C., Deneubourg, J. L., Blanco, S., Fournier, R., Jost, C. and Theraulaz, G. (2005). Self-organized aggregation in cockroaches. *Anim. Behav.* **69**, 169–180.

Judd, T. M. (1995). The waggle dance of the honeybee: which bees following a dancer successfully acquire the information? *J. Insect. Behav.* **8**, 343–353.

Kaib, M., Bruinsma, O. and Leuthold, R. H. (1982). Trail-following pheromone in termites, evidence for a multicomponent system. *J. Chem. Ecol.* **8**, 1193–1205.

Krause, J. and Ruxton, G. D. (2002). *Living in Groups*. Oxford: Oxford University Press.

Krebs, J. R. and Davies, N. B. (1984). *Behavioural Ecology: An Evolutionary Approach*, 2nd ed. Sunderland, MA: Sinauer.

Kummer, H. (1968). *Social Organisation of Hamadryas Baboons*. Chicago: University of Chicago Press.

Le Breton, J. and Fourcassié, V. (2004). Information transfer during recruitment in the ant *Lasius niger* L. (Hymenoptera: Formicidae). *Behav. Ecol. Sociobiol.* **55**, 242–250.

Levings, S. C. and Traniello, J. F. A. (1981). Territoriality, nest dispersion, and community structure in ants. *Psyche* **88**, 265–319.

Liefke, C., Holldobler, B. and Maschwitz, U. (2001). Recruitment behavior in the ant genus *Polyrhachis* (Hymenoptera, Formicidae). *J. Insect. Behav.* **14**, 637–657.

Lindauer, M. and Kerr, W. E. (1958). Die gegenseitige Verständigung bei den stachellosen Bienen. *Z. Vergl. Physiol.* **41**, 405–434.

Lindauer, M. and Kerr, W. E. (1960). Communication between workers of stingless bees. *Bee world* **41**, 29–41, 65–71.

List, C. (2004). Democracy in animal groups: a political science perspective. *Trends Ecol. Evol.* **19**, 168–169.

Lopez, F., Acosta, F. J. and Serrano, J. M. (1994). Guerilla vs. phalanx strategies of resource capture: growth and structural plasticity in the trunk trail system of the harvester ant *Messor barbarus*. *J. Anim. Ecol.* **63**, 127–138.

Lorenzo, M. G. and Lazzari, C. R. (1996). The spatial pattern of defaecation in *Triatoma infestans* and the role of faeces as a chemical mark of the refuge. *J. Insect. Physiol.* **42**, 903–907.

Lumsden, C. J. and Hölldobler, B. (1983). Ritualized combat and inter colony communication in ants. *J. Theor. Biol.* **100**, 81–98.

Mailleux, A. C., Deneubourg, J. L. and Detrain, C. (2000). How do ants assess food volume? *Anim. Behav.* **59**, 1061–1069.

Mailleux, A. C., Deneubourg, J. L. and Detrain, C. (2003a). How does colony growth influence communication in ants? *Insect. Soc.* **50**, 24–31.

Mailleux, A. C., Deneubourg, J. L. and Detrain, C. (2003b). Regulation of ants' foraging to resource productivity. *Proc. R. Soc. B Biol. Sci.* **270**, 1609–1616.

Mailleux, A. C., Detrain, C. and Deneubourg, J. L. (2005). Triggering and persistence of trail-laying in foragers of the ant *Lasius niger*. *J. Insect Physiol.* **51**, 297–304.

Mailleux, A.-C., Detrain, C. and Deneubourg, J. L. (2006). Starvation drives a threshold triggering communication. *J. Exp. Biol.* **209**, 4224–4229.

Michener, C. (1974). *The Social Behaviour of the Bees*. Cambridge, MA: Harvard University Press.

Miller, M. B. and Bassler, B. L. (2001). Quorum sensing in bacteria. *Annu. Rev. Microbiol.* **35**, 165–199.

Miller, D. M. and Koehler, P. G. (2000). Novel extraction of German cockroach (Dictyoptera: Blattellidae) fecal pellets enhances efficacy of spray formulation insecticides. *J. Econ. Entomol.* **93**, 107–111.

Millor, J., Pham-Delegue, M., Deneubourg, J. L. and Camazine, S. (1999). Self-organized defensive behavior in honeybees. *PNAS* **96**, 12611–12615.

Möglich, M., Maschwitz, U. and Hölldobler, B. (1974). Tandem calling a new kind of signal in ant communication. *Science* **186**, 1046–1047.

Nicolis, S. C. and Deneubourg, J. L. (1999). Emerging patterns and food recruitment in ants: an analytical study. *J. Theor. Biol.* **198**, 575–592.

Nicolis, G. and Prigogine, I. (1977). *Self-Organization in Non-Equilibrium Systems*. New York: Wiley.

Nicolis, S. C., Theraulaz, G. and Deneubourg, J. L. (2005). The effect of aggregates on interaction rate in ant colonies. *Anim. Behav.* **69**, 535–540.

Nieh, J. C., Contrera, F. A. and Nogueira-Neto, P. (2003). Pulsed mass recruitment by a stingless bee, *Trigona hyalinata*. *P. Roy. Soc. B Biol. Sci.* **270**, 2191–2196.

Oster, G. F. and Wilson, E. O. (1978). *Caste and Ecology in the Social Insects*. Princeton: Princeton University Press.

Pasteels, J. M. and Bordereau, C. (1998). Releaser pheromones in termites. In: *Pheromone Communication in Social Insects* (eds Vander Meer, R., Breed, M., Espelie, K. E. and Winston, M. L.), pp. 193–215. Boulder: Westview Press.

Pasteels, J. M., Deneubourg, J. L., Verhaeghe, J. C., Boevé, J. L. and Quinet, Y. (1986). Orientation along terrestrial trails by ants. In: *Mechanisms in Insect Olfaction* (eds Payne, T. L., Birch, M. C. and Kennedy, C. E. J.), pp. 131–138. Oxford: Oxford University Press.

Pasteels, J. M., Deneubourg, J. L. and Goss, S. (1987). Self-organization mechanisms in ant societies. (I). In: *Trail Recruitment to Newly Discovered Food Sources* (eds Pasteels, J. M. and Deneubourg, J. L.), pp. 155–175. From individual to collective behavior in social insects. Bâle: Experientia, Supplementum 54.

Perusse, D. and Lefebvre, L. (1985). Grouped sequential exploitation of food patches in a flock feeder, the feral pigeon. *Behav. Processes* **11**, 39–52.

Pontin, A. J. (1961). Population stabilization and competition between the ants *Lasius flavus* (F.) and *Lasius niger* (L.). *J. Anim. Ecol.* **30**, 37–54.

Portha, S., Deneubourg, J. L. and Detrain, C. (2004). How food type and brood influence foraging decisions of *Lasius niger* scouts. *Anim. Behav.* **68**, 115–122.

Pratt, S. C. and Sumpter, D. J. T. (2006). A tunable algorithm for collective decision-making. *Proc. Natl. Acad. Sci. USA* **103**, 15906–15910.

Pratt, S. C., Mallon, E. B., Sumpter, D. J. T. and Franks, N. R. (2002). Quorum sensing, recruitment, and collective decision-making during colony emigration by the ant *Leptothorax albipennis*. *Behav. Ecol. Sociobiol.* **52**, 117–127.

Quinet, Y. and Pasteels, J. M. (1991). Spatiotemporal evolution of the trail network in *Lasius fuliginosus* (Hymenoptera, Formicidae). *Belgian J. Theor. Biol. Zool.* **121**, 55–72.

Quinet, Y., de Biseau, J. C. and Pasteels, J. M. (1997). Food recruitment as a component of the trunk-trail foraging behaviour of *Lasius fuliginosus* (Hymenoptera: Formicidae). *Behav. Processes* **40**, 75–83.

Ratnieks, F. L. W. and Anderson, C. (1999). Task partitioning in insect societies. *Insect. Soc.* **46**, 95–108.

Ravary, F., Lecoutey, E., Kaminski, G., Chaline, N. and Jaisson, P. (2007). Individual experience alone can generate lasting division of labor in ants. *Curr. Biol.* **17**, 1308–1312.

Reinhard, J. and Kaib, M. (1995). Interaction of pheromones during food exploitation by the termites *Scherdorhinotermes lamanianu*. *Physiol. Entomol.* **20**, 266–272.

Reinhard, J. and Kaib, M. (2001). Trail communication during foraging and recruitment in the subterranean termite *Reticulitermes santonensis* De Feytaud (Isoptera, Rhinotermitidae). *J. Insect. Behav.* **2**, 157–171.

Richardson, T. O., Houston, A. I. and Franks, N. R. (2007). Teaching with evaluation in ants. *Curr. Biol.* **17**, 1520–1526.

Riley, J. R., Greggers, U., Smith, A. D., Reynolds, D. R. and Menzel, R. (2005). The flight paths of honeybees recruited by the waggle dance. *Nature* **435**, 205–207.

Rissing, S. W. and Wheeler, J. (1976). Foraging responses of *Veromessor pergandei* to changes in seed production (Hymenoptera-Formicidae). *Pan-Pac. Entomol.* **52**, 63–72.

Robinson, E. J. H., Jackson, D. E., Holcombe, M. and Ratnieks, F. L. W. (2005). Insect communication – 'No entry' signal in ant foraging. *Nature* **438**, p. 432.

Roces, F. (1990). Olfactory conditioning during the recruitment process in a leaf-cutting ant. *Oecologia* **83**, 261–262.

Roces, F. and Nunez, J. A. (1993). Information about food quality influences load-size selection in recruited leaf-cutting ants. *Anim. Behav.* **45**, 135–143.

Ruf, C., Costa, J. T. and Fiedler, K. (2001). Trail-based communication in social caterpillars of *Eriogaster lanestris* (Lepidoptera: Lasiocampidae). *J. Insect. Behav.* **14**, 231–245.

Saffre, F., Furey, R., Krafft, B. and Deneubourg, J. L. (1999). Collective decision-making in social spiders: dragline-mediated amplification process acts as a recruitment mechanism. *J. Theor. Biol.* **198**, 507–517.

Said, I., Costagliola, G., Leoncini, I. and Rivault, C. (2005)a). Cuticular hydrocarbon profiles and aggregation in four *Periplaneta species* (Insecta: Dictyoptera). *J. Insect. Physiol.* **51**, 995–1003.

Sakata, H. and Katayama, N. (2001). Ant defence system: a mechanism organizing individual responses into efficient collective behavior. *Ecol. Res.* **16**, 395–403.

Schmidt, V. M., Schorkopf, D. L., Hrncir, M., Zucchi, R. and Barth, F. G. (2006a). Collective foraging in a stingless bee: dependence on food profitability and sequence of discovery. *Anim. Behav.* **72**, 1309–1317.

Schmidt, V. M., Zucchi, R. and Barth, F. G. (2006b). Recruitment in a scent trail laying stingless bee (*Scaptotrigona depilis*): changes with reduction but not with increase of energy gain. *Apidologie* **37**, 487–500.

Schorkopf, D. L., Jarau, S., Francke, W., Twele, R., Zucchi, R., Hrncir, M., Schmidt, V., Ayasse, M. and Barth, F. G. (2007). Spitting out information: *Trigona* bees deposit saliva to signal resource locations. *Proc. R. Soc. B Biol. Sci.* **274**, 895–898.

Seeley, T. D. (1992). The tremble dance of the honeybee: message and meanings. *Behav. Ecol. Sociobiol.* **31**, 375–383.

Seeley, T. D. (1994). Honey bee foragers as sensory units of their colonies. *Behav. Ecol. Sociobiol.* **34**, 51–62.

Seeley, T. D. (1995). *The Wisdom of the Hive*. Cambridge: Harvard University Press.

Seeley, T. D. and Tovey, C. A. (1994). Why search time to find a foodstorer bee accurately indicates the relative rates of nectar collecting and nectar processing in honeybee colonies. *Anim. Behav.* **47**, 311–316.

Seeley, T. D. and Towne, W. F. (1992). Tactics of dance choice in honey-bees – do foragers compare dances? *Behav. Ecol. Sociobiol.* **30**, 59–69.

Seeley, T. D. and Visscher, P. K. (1985). Survival of honeybees in cold climates: the critical timing of colony growth and reproduction. *Ecol. Entomol.* **10**, 81–88.

Seeley, T. D. and Visscher, P. K. (2003). Choosing a home: how the scouts in a honey bee swarm perceive the completion of their group decision making. *Behav. Ecol. Sociobiol.* **54**, 511–520.

Seeley, T. D. and Visscher, P. K. (2004). Quorum sensing during nest-site selection by honeybee swarms. *Behav. Ecol. Sociobiol.* **56**, 594–601.

Seeley, T. D., Camazine, S. and Sneyd, J. (1991). Collective decision-making in honey-bees – how colonies choose among nectar sources. *Behav. Ecol. Sociobiol.* **28**, 277–290.

Seeley, T. D., Kühnholz, S. and Weidenmüller, A. (1996). The honeybee's tremble dance stimulates additional bees to function as nectar receivers. *Behav. Ecol. Sociobiol.* **39**, 419–427.

Seeley, T. D., Mikheyev, A. S. and Pagano, G. P. (2000). Dancing bee tunes both duration and rate of waggle-run production in relation to nectar-source profitability. *J. Comp. Physiol. A* **186**, 813–819.

Seeley, T. D., Visscher, P. K. and Passino, K. M. (2006). Group decision making in honey bee swarms. *Am. Scientist* **94**, 220–229.

Sendova-Franks, A. B. and Franks, N. R. (1999). Self-assembly, self-organization and division of labour. *Philos. Trans. R. Soc. B Biol. Sci.* **354**, 1395–1405.

Sherman, G. and Visscher, P. K. (2002). Honeybee colonies achieve fitness through dancing. *Nature* **419**, 920–922.

Shirakawa, T. and Gunji, Y. P. (2007). Emergence of morphological order in the network formation of *Physarum polycephalum*. *Biophys. Chem.* **128**, 253–260.

Sørensen, A. A., Busch, T. M. and Vinson, S. B. (1985). Control of food influx by temporal subcastes in the fire ant, *Solenopsis invicta*. *Behav. Ecol. Sociobiol.* **17**, 191–198.

Stickland, T. R., Britton, N. F. and Franks, N. R. (1999). Models of information flow in ant foraging: the benefits of both attractive and repulsive signals. In: *Information Processing in Social Insects* (eds Detrain, C., Deneubourg, J. L. and Pasteels, J. M.), pp. 83–100. Basel: Birkhaüser Verlag.

Sumpter, D. J. T. (2006). The principles of collective animal behaviour. *Philos. R. Soc. B* **361**, 5–22.

Sumpter, D. J. T. and Beekman, M. (2003). From nonlinearity to optimality: pheromone trail foraging by ants. *Anim. Behav.* **66**, 273–280.

Sumpter, D. J. T. and Pratt, S. C. (2003). A modelling framework for understanding social insect foraging. *Behav. Ecol. Sociobiol.* **53**, 131–144.

Takahashi, K. H. (2006). Spatial aggregation and association in different resource-patch distributions: experimental analysis with *Drosophila*. *J. Anim. Ecol.* **75**, 266–273.

Tanner, D. A. and Visscher, K. (2006). Do honey bees tune error in their dances in nectar-foraging and house-hunting? *Behav. Ecol. Sociobiol.* **59**, 571–576.

Tautz, J. and Sandeman, D. C. (2003). Recruitment of honeybees to non-scented food sources. *J. Comp. Physiol. A* **189**, 293–300.

Tautz, J., Casas, J. and Sandeman, D. (2001). Phase reversal of vibratory signals in honeycomb may assist dancing honeybees to attract their audience. *J. Exp. Biol.* **204**, 3737–3746.

Theraulaz, G. and Spitz, F. (1997). *Auto-organisation et comportement*. Paris: Hermès.

Thomas, J. and Valone, T. J. (1989). Group foraging, public information, and patch estimation. *Oikos* **56**, 357–363.

Towne, W. F. and Gould, J. L. (1988). The spatial precision of the honey bees dance communication. *J. Insect. Behav.* **1**, 129–156.

Traniello, J. F. A. (1982). Recruitment and orientation components in a termite trail pheromone. *Naturwissenschaften* **69**, 343–344.

Traniello, J. F. A. and Leuthold, R. (2000). Behavior and ecology of foraging in termites. In: *Termites: Evolution, Eusociality, Symbioses, Ecology* (eds Abe, T., Bignell, D. E. and Higashi, M.), pp. 141–168. Dordrecht, The Netherlands: Kluwer.

Traniello, J. F. A. and Robson, S. K. (1995). Trail and territorial communication in social insects. In: *Chemical Ecology of Insects*, Vol. 2 (eds Bell, W. J. and Cardé, R.), pp. 241–286. New York: Chapman and Hall.

Traniello, J. F. A., Fourcassié, V. and Graham, T. P. (1991). Search behavior and foraging ecology of the ant *Formica schaufussi* – colony-level and individual patterns. *Ethol. Ecol. Evol.* **3**, 35–47.

Valone, T. J. and Templeton, J. J. (2002). Public information for the assessment of quality: a widespread social phenomenon. *Philos. Trans. R. Soc. B Biol. Sci.* **357**, 1549–1557.

Verhaeghe, J. C. (1982). Food recruitment in *Teramorium impurum* (Hymenoptera, Formicidae). *Insect. Soc.* **29**, 67–85.

Verhaeghe, J. C. and Deneubourg, J. L. (1983). Experimental study and modelling of food recruitment in the ant *Tetramorium impurum* (Hym. Form.). *Insect. Soc.* **30**, 347–360.

Visscher, P. K. (2007). Group decision making in nest-site selection among social insects. *Annu. Rev. Entomol.* **52**, 255–275.

Visscher, P. K. and Camazine, S. (1999a). Collective decisions and cognition in bees. *Nature* **397**, p. 400.

Visscher, P. K. and Camazine, S. (1999b). The mystery of swarming honeybees: from individual behaviors to collective decisions. In: *Information processing in social insects* (eds Detrain, C., Deneubourg, J. L. and Pasteels, J. M.), pp. 355–378. Basel: Birkhaüser Verlag.

Vittori, K., Talbot, G., Gautrais, J., Fourcassie, V., Araujo, A. F. R. and Theraulaz, G. (2006). Path efficiency of ant foraging trails in an artificial network. *J. Theor. Biol.* **239**, 507–515.

Von Frisch, K. (1967). *The dance language and orientation of bees.* Cambridge: Harvard University Press.

Watmough, J. and Edelstein-Keshet, L. (1995). Modeling the formation of trail networks by foraging ants. *J. Theor. Biol.* **176**, 357–371.

Weidenmuller, A. and Seeley, T. D. (1999). Imprecision in waggle dances of the honeybee (*Apis mellifera*) for nearby food sources: error or adaptation? *Behav. Ecol. Sociobiol.* **46**, 190–199.

Wertheim, B., van Baalen, E. J. A., Dicke, M. and Vet, L. E. M. (2005). Pheromone-mediated aggregation in nonsocial arthropods: An evolutionary ecological perspective. *Annu. Rev. Entomol.* **50**, 321–346.

Wilson, E. O. (1962a). Chemical communication among workers of the ant *Solenopsis saevissima* (Fr. Smith). 2. An information analysis of the odour trail. *Anim. Behav.* **10**, 148–158.

Wilson, E. O. (1962b). Chemical communication among workers of the ant *Solenopsis saevissima* (Fr. Smith). 3. The experimental induction of social responses. *Anim. Behav.* **10**, 159–164.

Yano, S. (2008). Collective and solitary behaviors of two-spotted spider mites (*Acari Tetranychus*) are induced by trail following. *Ann. Entomol. Soc. Am.* **101**, 247–252.

Index

CPI Antony Rowe
Eastbourne, UK
June 08, 2020

Monday, September 15th. The regiment commenced to march quite early in the morning, and passed through Fairfield and Middletown. We could here begin to form some idea of that great army, the "Army of the Potomac," and the fearful destruction that an army can make. The road was completely blocked up with army wagons and ambulances. The road was narrow over the mountain, and terribly dusty. The ambulances were filled with the wounded, and rebel prisoners under guard were trying to go to the rear. Infantry, baggage wagons, provision and ammunition trains, were eagerly pushing to the front. The result was a stand-still for over an hour. On both sides of the road, shot and shell had pierced the trees and houses. The fences were riddled with bullets, telegraph poles were down, and the earth was ploughed by solid shot. The dead lay by the road-side, and the ambulances were scouring the mountain sides with men detailed to pick up the wounded. The churches, houses, and barns were filled with the wounded. Parties were seen in every direction burying the dead. The scenes showed that a fierce battle had been fought the day before, and we began to realize what we must go through when we should join the main army. We marched that day about twelve miles, and encamped for the night on the battle-field of South Mountain. The next day we started on the march at six A.M. and passed through Boonsboro, and Keedysville. At Boonsboro, also, the churches, houses, and barns were filled with wounded. At Keedysville, we had a long rest, and it was here that we first saw a "line of battle." Colonel Beach, with his experienced eye, first spied the distant jets of white smoke. All were watching the peculiar puffs of smoke with great interest, when Adjutant Burnham, who had been absent, returned with the order that we were wanted at the front. This took us a little by surprise as we did not expect to go into battle so soon. But on went the bundles, and after a tedious march through ploughed fields and forests, passing brigades and divisions, the booming of artillery and bursting of shells sounding louder and louder, we finally joined a brigade consisting of the 4th R.I., and the 8th and 11th C.V.

After resting awhile we loaded our muskets for the first time, and marched over a hill, and into a meadow which lay between two hills. While getting into this position we could plainly see the rebel gunners load and fire, some of the shells coming quite near us. At last we were in the great "line of battle" of the "Army of the Potomac," 2d Brigade, 3d Division, 9th Corps, General Burnside, on its extreme left. It was now eight o'clock in the evening, and quite dark; we were within a few rods of the enemy, and orders were given in a whisper; we were ordered to make no noise and to rest on our arms; for thirty minutes the utmost quiet prevailed. A musket was accidentally discharged; in a second the troops were on their feet, with arms at a "ready," and as they stood peering into the darkness ahead you could hear both lines of battle spring to arms for miles.

Occasionally the boom of artillery was heard, and during the night there were repeated alarms, so that the soldiers on either side obtained but little rest. The hostile pickets on one portion of the line were so near each other, that during the night six of the enemy were captured.

CHAPTER II.

1862.

THE BATTLE OF ANTIETAM.

The next morning dawned beautifully; little did we imagine that that bright sun would be obscured by the smoke of battle, the field we trod ploughed with shot, flow with blood, and planted thick with the dead. Scarcely had the sun risen when a shell from the enemy dropped not far from our force, which was quietly resting upon their arms near the crest of a low knoll a short distance from the enemy's position. Immediately another followed, a twelve pounder crashed diagonally through the Eighth Connecticut, killing three men instantly, and wounding four in Company D.

The position was changed for one less exposed, but in getting there the troops were obliged to pass under a deadly fire from a rebel battery stationed at short range distance. In this undertaking the Sixteenth lost three wounded. We lay here perhaps two hours, and had a good view of the battle on the right, which had by this time assumed a fearful magnitude. Along the western banks of the Antietam River, there runs, with a gradual rise of undulating ground, a crescent-shaped ridge, presenting its concave side to the river. The top of this ridge spreads out into a broad tableground of forests and ravines. A series of timbered-covered hills surrounded this ridge; some of the adjacent hills had been cleared of the forest, and were covered with orchards and cornfields, enclosed with fences of rails or stone. Behind this ridge runs the road from Hagerstown to Sharpsburg and Shepardstown. Sharpsburg is just in the rear of the ridge.

Along these hills the rebel lines were posted, four miles in extent. Their position was exceedingly strong, protected by ravines and forests. Every commanding crest bristled with artillery, and the forests were planted thick with infantry. The extreme right of the rebel line was within three-fourths of a mile of the Potomac; in front, and along their left flank, flowed the Antietam, winding through a wooded ravine, with banks too high and with waters too deep to permit a crossing, except at two fords, at some distance from each other. Between these distant fords there were three bridges; on the right, at the center, and on the left. These bridges were strongly guarded. The federal troops were on the east side of the Antietam, behind a low range of hills, lying at the base of the Blue Ridge. These eminences were generally commanded by the heights held by the rebels. General Lee had certainly chosen a very strong position.

The Eleventh Connecticut now received orders from General Burnside to take the bridge, after the batteries had shelled the woods on the other side, and hold it until General Rodman could march his column over. At about nine o'clock the Sixteenth again formed and marched about a mile, first through a corn-field, and finally into a valley where they halted in an orchard. While passing through the cornfield the men stripped themselves of blankets, overcoats, and all luggage that would impede the progress of marching or the use of firearms. After filling our canteens from a brook near by, we marched up a steep hill that seemed almost impossible to surmount, then down on the other side and into Antietam river, which we forded and marched to a side hill. Soon in plain sight could be seen a rebel battery dashing intrepidly forward and planting itself directly in range of the Sixteenth.

By this time the rebel batteries were all roaring. They opened on us in all their fury. The air was filled with bullets and fiendish missiles. Hundreds of cannon were now aimed at us; grape and cannister, marbles and railroad iron were showered down like rain. The crest of the hill was a great protection to the Sixteenth, and only about a dozen were disabled. A battery was ordered up to engage the enemy, but it was whirled back in less than five minutes, losing every officer, seven men, and five horses. To see those men stand there and be shot down till they received orders to retire was a fearful sight. It was half past three o'clock; the Fourth Rhode Island and the Sixteenth Connecticut were ordered into a cornfield, and they moved forward quite a distance in advance of the army at their right; we here laid down letting the shot and shell pass over us.

In the meanwhile the Division of A.P. Hill, which had arrived from Harper's Ferry, and joined Lee's army, were coming into this cornfield from the opposite side, unobserved; at the same time Company H, (Captain Barber,) had been thrown out in advance as a vidette to prevent being surprised. At four o'clock McClellan sent orders to Burnside to advance, and carry the batteries in his front at all hazards and at any cost. Burnside's corps was charging. General Rodman observed that the rebels were about to flank us and get in our rear, and ordered the Fourth Rhode Island, and Sixteenth Connecticut to swing to the left that we might face them, but at that particular moment the rustling of cornstalks warned us that the rebels were on us. Colonel Beach gave the order 'Attention!' While this order was being executed a terrible volley was fired into us. Volley after volley in quick succession was hurled into our midst. The Sixteenth sprang up and returned the fire with good effect; some fixed bayonets, advanced, and were captured. The most helpless confusion ensued. Our men fell by scores on every side. Still our position was obstinately maintained, until ordered to fall back. The rebels discovered the disorder, and came on us in heavy column.

While we were falling back to cover near the bridge we were swept by a destructive cross-fire, and the rebels becoming entangled in this cross-fire extricated themselves and fell back to the stone wall. The Eighth, Eleventh, and Sixteenth Connecticut, and the Fourth Rhode Island, re-formed and were placed in position for defence. At this time General Burnside's messenger rode up to McClellan. His message was, "I want troops and guns. If you do not send them I cannot hold my position for half an hour." McClellan said slowly: "Tell General Burnside that this is the battle of the war. He must hold his ground till dark at any cost. I will send him Miller's Battery; I can do no more. I have no infantry." Then as the messenger was riding away he called him back. "Tell him if he cannot hold his ground, then the bridge, to the last man! always the bridge! If the bridge is lost, all is lost." The enemy was pressing down hard upon the battery which had been placed on the crest in front of the Eleventh. Burnside called for aid and General Rodman having been killed, Colonel Harland took command of the division, re-formed the disorganized regiments, and by his bravery the unsupported battery was rescued from capture.

The fighting was ended. It was indeed a fearful day for the Sixteenth. Without having time allowed to learn even the rudiments of military science, it was hurried forward and was formed in regimental line almost for the first time on the battle-field of Antietam, the bloodiest day America ever saw. After sunset the brigade was relieved, and retired across the river to reorganize and be ready for the duties which they might be called upon to do when another day should come. Arms were stacked, and the tired soldiers laid down to rest. Of all gloomy nights, this was the saddest we ever experienced. All was quiet and silent as the grave. The stacks of straw which the rebels had fired burned slow and dimly. The cries and groans of the wounded that lay on the battle-field could be heard distinctly, and the occasional report of artillery sounded solemn and death-like.

The morning of the 18th dawned. The sun rose obscurely and there was a fair prospect of rain. The Sixteenth had gone into the battle with 940 men. Some not being able to endure the hard marching had been left at Frederick City. On this morning we could muster but 300 men; but during the day about 200 joined the regiment who wore unable to find it the previous evening. It was a sorry sight that morning as General Burnside rode up to encourage the men, who supposed, of course, that the battle would be resumed, said, "only hold out this day, boys, and the war is ended." Colonel Harland's brigade was once more moved forward, and stationed in line of battle near the bridge, which General Burnside had been ordered to hold at all hazards. Here they remained until the next morning, when the bridge was crossed, and the Sixteenth detached from the brigade to bury their dead, and care for the wounded who were still lying upon the field. The casualties in the Sixteenth were as follows: Lieut. Col. F.W. Cheney wounded in the arm, Maj. Geo. A. Washburn wounded severely in the groin, Captains Manross, Drake, and Brown instantly killed, Captain Barber mortally wounded, dying; about fifteen hours after, Captains Babcock and Hayden wounded, Lieut. William Horton killed, and four lieutenants wounded.

Thirty-eight enlisted men were killed outright. A great many of the men were mortally wounded and died within twenty-four hours after the battle, so that on the 19th, two days after the battle, when the regiment was detailed to bury the dead and pick up the wounded, the recapitulation stood as follows: Killed, 4 captains, 1 lieutenant, and 51 enlisted men; wounded, 2 field officers, 2 captains, 4 lieutenants, and 176 enlisted men; captured, 12 enlisted men and 180 missing, making a total loss of 432 men. For forty-eight hours men were brought in. Parties scoured the fields hunting for the wounded. Many had crept out of the storm of battle and hidden under fences, or among rocks, or in thickets, and their strength failing, they could neither come forth, or make known their situation. Some of the badly wounded did not have any attention for several days. All houses and barns were converted into hospitals, and yards and fields were strewn with straw and the wounded laid, there without shelter. Surgeons worked hard day and night, taking rest only when unable to stand up from weariness. At one of these hospitals about 25 of the Sixteenth were placed. Nothing was to be heard but cries, groans, and entreaties. Here Captain Barber lay in about the center of a barn, quiet, happy, and contented with his lot. The wounded lay around him on every side. He said that he could not live long, and spoke encouraging words to all. Gilbert B. Foster, of Co. A, who died November 13th, was also here. In a room about 12×20 a bloody table stood and around it were five surgeons. A wounded man was laid on the table and it took but a few seconds for them to decide what to do, and but a few minutes to do it. The amputated limbs were thrown out of a window. In forty-eight hours there were as many as two cart loads of amputated legs, feet, arms, and hands in the pile. Plenty of men, most of them slightly wounded, were hard at work carrying the wounded to and fro, making beds of straw, hauling and cutting wood, cooking, feeding, and assisting in a thousand ways.

(On the afternoon of the 18th, a heavy shower, lasting an hour, made it very uncomfortable for those not sheltered.)

"Captain Drake was the most gentlemanly man in the regiment," said Surgeon Mayer. "He was the very soul of courtesy and unaffected dignity of deportment. He always had a quiet care for his men, when they were sick, and was a marked favorite with them, as well as with comrades in the line."

"Capt. Barber was especially noticeable for his religious character, earnest convictions, and high regard for duty. His patriotism was of sterling mould, and he was a brave and intelligent officer."

"Captain N.S. Manross, of Bristol, was a man of learning and varied accomplishments. He graduated at Yale in the class of 1850. In 1861, Dr. Manross accepted the position of Professor of Chemistry and Botany in Amherst College, where he was very popular and successful. Previous to this he had been to Europe, attended German lectures, and took the degree of doctor of philosophy. He invented a machine for the cutting of crystals from calc-spar. During vacation, he returned to Bristol, Conn., where he made a patriotic speech to his fellow-citizens, and consented to lead them to the field. Said he to his wife, "You can better afford to have a country without a husband than a husband without a country." His men loved him. While the regiment was in the cornfield and the baffle was raging the fiercest, a cannon-ball struck Captain Manross in the side and passed under his arm. A friend bending over him heard him murmuring, "Oh, my poor wife, my poor wife!"

Prof. James D. Dana said of him, "His death is a great loss to the scientific world." Prof. B. Silliman, Jr., says "As an explorer, Dr. Manross possessed remarkable qualifications. To a rugged constitution and great powers of endurance, he united great coolness, quiet but undaunted demeanor, the courage of a hero, and unyielding perseverance. Had he lived—but what need is there of conjecture now? The world will never know its loss, but his friends will never forget theirs."[1]

---

[1] Military and Civil History of Connecticut.

On the 19th, the Sixteenth were employed in gathering up the dead and wounded. This was a very unpleasant duty, making many of the men sick. Forty of the men were buried that afternoon side by side, under a large tree, near the stonewall, where the hardest of the battle was fought.

BELINDA SPRINGS, ANTIETAM IRON WORKS, AND PLEASANT VALLEY.

The following day the regiment rejoined their brigade at Belinda Springs, a distance of two miles, and moved thence to Antietam Iron Works on the 26th. Here sickness prevailed to a great extent, and but few men could be reported for duty. On the afternoon of September 23d, Messrs. E.N. Kellogg, J.M.B. McNary and W.H.D. Callender, of Hartford, Conn., came into camp. Crowds gathered around them, anxious to learn the news from home, and to send letters and messages. It seems that at 10 P.M., Saturday, Sept. 20th, a dispatch was received at Hartford, that the Sixteenth had suffered severely and that Lieut. Col. Cheney was killed. It was thought best that these gentlemen should proceed to the battle-field, and carry out such arrangements for the care of the dead and wounded as they thought best; they accordingly left Hartford on the midnight train, reaching the regiment as stated above. By this time the dead were all buried, and most of the wounded had been taken to various hospitals. Lieut. Col. Cheney and Major Washburn were at this time at Boonsboro.

On October 3d, the entire army was reviewed by President Lincoln. The Vice-President and several Congressmen were present. On October 7th, the regiment marched over the mountain into Pleasant Valley, a distance of about six miles. This march, though short, was a very hard one; the path being very narrow, only admitting one at a time in some places, and so steep and rocky that it was very hard to surmount with our heavy loads. At the top of the mountain the troops halted an hour for rest. Here we had an extended view of the surrounding country. We could see a great distance, and the scenery was magnificent. At the camp in Pleasant Valley the regiment suffered severely from sickness, and when the army again took up its line of march, they could muster but few effective men. At this place a large number of promotions and appointments were made to fill vacancies.

MARCH TO FALMOUTH.

On October 28th we struck tents at 8 A.M. and after "falling in" we were once more "on the march." We passed through Knoxville and Berlin. At Berlin we crossed the Potomac on a pontoon bridge. Once more we trod the "sacred soil" of Virginia. Passing through Lovettsville, we halted at 2 P.M. and encamped about a mile from the village. On October 30th reveille was sounded at 3 A.M. By the time the men had struck tents and packed up, the cooks had plenty of hot coffee ready, which is the soldier's breakfast, and at sunrise we were again on the march. We passed through a village called Burlington and encamped at 11 A.M. near Wheatland. Saturday, November 1st, there was heavy firing in front during the afternoon. Orders were given to be ready to march at a moment's notice. The next day we began to march at 9 A.M. with five days rations. We passed through Princeville and Goose Creek. The heavy firing in front continued. We halted at 7 P.M. and went into camp. The next day we marched during the afternoon, passing through Union. Artillery firing was heard ahead. On November 5th we struck tents and were on the march at 8 A.M. At 2 P.M. the entire army was drawn up in line of battle about a mile beyond Rectorsville, the artillery doing the fighting. After resting on our arms all night we commenced to march at 8 A.M., making fifteen miles during the day. We were following the enemy up closely. November 7th was a tedious and rough day. Snow fell most of the day, and at least one-third of the regiment were without shoes. We marched to a place called Waterloo, within five miles of Warrenton. Colonel Beach, being absent sick, Lieut. Col. Cheney and Maj. Washburn wounded, Capt. Mix was in command. On the 9th of November some Rebel cavalry broke through our lines and were making a raid around the army. Coming very near us at 4 P.M. the long roll beat and without packing up and hardly having time to put on our equipments, we double-quicked up the mountain and took position in ambush, where we staid till the morning of the 11th, when we returned to the old camp.

For several days rations had been very scarce, hard crackers selling as high as twenty-five cents each. Rations of pork, beans, and potatoes finally came on the 11th, and the next day some hard bread, which was very wormy. Rations not being plenty, the men went foraging, and obtained large quantities of honey. One man who was detailed in the Quartermaster's Department, who always had considerable lip, was successful enough to get two water-pails full. While eating some in the night he was stung by a bee, and the next morning he had about four inches of lip, which was rather more than we had seen him have before.

On November 15th, we commenced to march at eight o'clock in the morning. After marching about three hours, we suddenly halted on the Warrenton turnpike, near Sulphur Springs. Here the cavalry and light batteries had a little set-to with the rebels, in which the rebels got worsted and retreated, leaving two wagons loaded with muskets, which they set on fire. After filing to the left through the woods, and into a hollow near the road, we drew up into line of battle for the night.

On Sunday, November 16th, we marched twenty miles, passing through the town of Liberty. The next day we marched from noon till eight in the evening, passing through Elk Run. On November 18th, the reveille was sounded at three A.M. We marched from six A.M. till five P.M., making a very long and hard march, and many of the men fell out from exhaustion.

On November 19th, the reveille sounded at five A.M., and at eight o'clock we were once more on the march. A hard rain-storm which had set in the night before made the road in this region from three to eight inches deep with mud. During the march we forded four streams, knee deep. We passed through Spottsville and Falmouth.

Falmouth, Camp Starvation.

We encamped opposite the city of Fredericksburg, at half past two in a drenching rain, having made a hundred and seventy-five miles in twelve days. The field where the regiment encamped was very even ground, and the water stood on it from half an inch to two inches deep. The mud was about four inches deep. The men were completely exhausted from scanty rations, and foot-sore from long marches. The rain coming down in torrents, the soldiers were wet through to the skin. Fires could not be built, and tents could not be raised. Little or no sleep did the troops get that night. The next day was very cold, and it was still raining. The batteries opened on the city for about two hours in the morning.

On November 21st, the sun once more showed itself after a long absence, and the men were enabled to dry their clothes, build fires, cook and eat salt junk, pour down hot coffee, and once more felt in good spirits.

On the 26th, General Sumner reviewed the entire corps.

On December 3d, Arthur D.N. Talcott, of Company "A," died in camp, and was buried at sundown.

When the regiment left camp near Fairfax Seminary on the 7th of September, they left their knapsacks with contents under charge of a guard. A few days after they were sent to Washington, and there stored. These were returned to us on the 3d of December. They were very welcome at this time, the weather now being very cold. The snow was three inches deep, and there was plenty of ice. For nearly three months a number of the men had been without blankets. About this time Governor Morgan of New York sent us a taste of home. Each man had three apples, two onions, and half a pickle, and the smoking men had half a paper of tobacco each. These went down with a genuine relish. At this time Capt. Charles

L. Upham, of the Eighth Connecticut Volunteers, was placed in command of the regiment.

CHAPTER III.

1862.

BATTLE OF FREDERICKSBURG.

On Wednesday, December 10th, clothing was issued to the regiment. Shoes were very much needed. In the evening a pontoon train went down towards the river, but no unusual notice or remarks were made about it, and both officers and men went to sleep that night without suspecting in the least that early on the morrow a heavy battle would be raging. The next morning the troops were early aroused by the tremendous discharge of two mortars, and simultaneously the opening of our batteries of nearly two hundred pieces. Nearly the entire day the batteries poured incessantly their deadly fire of shot and shell into the city with terrible rapidity. During the afternoon the firing gradually ceased, and at sundown victory rested on our banners. During the day three days rations and sixty rounds of cartridges were issued to the men. Towards the evening the Sixteenth was ordered down to the river, but before reaching there the order was countermanded, and they returned to camp for the night. The next day (Friday,) the Sixteenth advanced to the river again early in the morning, and lay on the banks all day, watching the fighting on the other side of the stream. In the evening they crossed the pontoon bridge, and went into the city. After stacking arms on Main street, most of the men went into houses to sleep. The effects of this short siege was awful to contemplate. Some portions of the city were completely battered down. Buildings in various parts of the city were burning, and during the night fresh fires were continually breaking out. Although the enemy had carried away most of their wounded and dead, still a few remained in the city. In a cellar was found by the Union troops, ten women and a child, all dead; they had gone there for protection from our shells, but one had struck there, and bursting, killed them all. While a member of the Sixteenth was searching for wood in the yard of a residence after dark, he stumbled over what he supposed to be soldiers asleep on the ground. Excusing himself he went on and after gathering an armful of wood, was returning when he stumbled over the same men again. Much to his disappointment they did not get up and damn him. Going into the house and getting a lighted brand, he came out and found that they were three dead rebels who had been killed and lay there side by side. One of them was an officer. An amusing incident occurred on this same evening in Company H. Sergeant Spencer was around the yard looking after boards to sleep on. Finding one that was some twelve or fourteen feet long, he laid one end of it on what he supposed to be a stone, and was about to jump on it to break it in the center, when a soldier who lay there wanted to know "what he was trying to do?" In the darkness of the night he had laid the board on a man's head. The next day we were drawn up in line of battle, but being on the reserve had nothing to do but witness the contest raging in front, which was fearful. At dusk we moved to the front, where bullets came thick and fast until eight o'clock, when the firing ceased, and all was quiet during the night, except the howling of dogs, and the occasional discharge of artillery.

On Sunday morning, December 14th, we returned into the city, remaining there all day. The fighting continued hard in front. At sundown we again moved to the front, where we remained supporting a battery until the next night, when we returned to the city, crossed the river and marched to our old camp, being the last brigade to leave the city. Thus ended the battle without the Sixteenth being actively engaged with the enemy, and meeting with a loss of only one wounded and one missing. Drilling, inspections, grand reviews, picket duty, and frequent preparations for marching, constituted the chief occupation of the troops during the greater part of the winter months. The weather was extremely cold, quarters were poor, and constant exposure invited sickness and disease, and death creeping in boldly hurried away its defenceless victims with alarming rapidity. Many were discharged, and the ranks continued to decrease daily.

On December 24th, Lieutenant-Colonel Cheney, in consequence of the severity of his wound was compelled to resign, and two days after, Adjutant John H. Burnham was promoted to be Lieutenant Colonel, and took command, Colonel Beach being absent, sick. Colonel Burnham's promotion was the making of the regiment. Being a man of promptness, and full of energy, and above all a perfect soldier, Colonel Burnham infused a new spirit into an organization which had been exhausted by arduous marches, a severe battle, and a weary campaign. It was about this time that another piece of good luck happened to the regiment, which was the appointment of Dr. Mayer, as Surgeon. He was a good physician, and as a surgeon could not be surpassed in the army of the Potomac. He commenced immediately to make improvements in and out of the hospital, and to look to the cleanliness of the tents, company streets, and the cooking utensils. He also saw that the food issued was properly prepared by the cooks; and when he gave cough syrup, it was not stuff that men would use on their food for molasses.

NEWPORT NEWS AND SUFFOLK.

On the 6th of February, 1863, our connection with the Army of the Potomac was dissolved. We were ordered to Newport News, where the regeneration of the regiment steadily progressed. At three o'clock in the morning the regiment was ordered to pack up, and be ready to march in two hours. It was pitch dark and raining terribly, with mud six inches deep. It was some little time before the men could get bon-fires burning, so that we could see to pack up. The men dressed, took what rations they could get, and fell into line after repeated orders, leaving most of the tents standing, they being wet and too heavy to carry. Marching to the depot, the mud was not only deep but extremely slippery, and nearly every man slipped down, and those that did not, were completely spattered over from head to foot, and were covered with mud and completely drenched through to the skin by the rain, which was decidedly uncomfortable that cold morning. After shivering in the cold for two hours, we were allowed to get aboard the freight cars, and were taken to Acquia Creek. Here we went aboard the steamer John S. Brooks, as did also the 8th and 15th C.V. Most of the men had little or no water in their canteens, and all suffered terribly from thirst before we reached Newport News on the afternoon of the 8th. Whose fault this was I am unable to say, but it was a great piece of negligence to put troops aboard a vessel knowing that they were to remain there for over two days without seeing them provided with water. Newport News was a paradise by the side of Falmouth. There was no mud, rations were good, and the weather was beautiful. We were quartered in barracks, which made it very pleasant.

On landing, the first thing was a drink of water, and then two days rations of soft bread were issued to us, which, although being two loaves, was disposed of in less than two hours; the first one in something less than five minutes. Not having tasted any for over five months we appreciated it. The guard mounts, dress parades, and reviews at this place were the grandest and most imposing ever witnessed in this country. About five weeks were spent in drilling, recruiting, &c., at the end of which time the 3d Division was ordered to Suffolk to strengthen the force at that point.

It was the early part of March and bitter cold when the regiment left Newport News and by boat went to Norfolk, where they went aboard a train consisting of platform cars with a single baggage or passenger car in the rear for the officers. Everything being in readiness the train started and sped on its way to Suffolk, arriving there at midnight. The men jumped off the train and fell into line, when lo and behold only two officers were to be found. The car containing the officers had not been attached to the train. The two officers present had for some reason got upon the cars with their men, and therefore were with the regiment. But we had no orders, whether we were to go farther or stay there, and whom to report to we did not know. The train moved off, dark as a pocket, and some of the men nearly froze to death. Something had got to be done. After a little consultation, the two officers took command of the two wings respectively, and the First Sergeants the companies. The regiment moved off by the flank to cut their way through the darkness and encamp somewhere, until daylight. We first tumbled down a steep embankment, at least twelve feet, the men falling on all sides, then into a brook two feet deep and six feet wide, and finally brought up against a rail fence. Tearing this down we passed into a field and halted, not deeming it best to proceed farther. The men spread out in every direction in the darkness, each one bringing in what he could find in the shape of wood to build fires.

There was a house near by which we supposed to be vacant, and the men in the darkness had taken all the fence and wood, and had even pulled the clapboards from the house as high as they could be reached. When morning came, we found it to be an elegant wood house painted white, and the owner thereof at first made quite a fuss, but when he found so many of the men nearly frozen to death, he concluded it was all for his country. It was on that night that the Quartermaster-Sergeant found that one of the men had bored a hole into a barrel of coffee, which he had mistaken for whiskey, and was shaking it up good, wondering why it would not run. Daylight finally came and we found that we were on the outskirts of the city and within sixty rods of the 112th N.Y. Vols., whose generous Colonel hearing the noise in the night, reconnoitered and finding that we were Union troops, ordered all his cooks up to make us hot coffee. Kettle after kettle of hot coffee all sweetened, was brought to us, which we drank in large quantities before getting thoroughly warmed through. This was a perfect godsend to us, and a more thoughtful action could not have been done by the Colonel. We fully appreciated it, as was shown by the fast friendship between the two regiments thereafter. Some half dozen of the men nearly died, by being chilled through, being several days before they were able to do duty. The officers arrived next morning on the regular train.

CHAPTER IV.

1863.

SIEGE OF SUFFOLK.

During the siege of Suffolk the Sixteenth took an active part on the defensive side, and had the honor of two engagements with the enemy, in one sally losing one killed and seven wounded, and in a sort of half battle across the Nansemond river, two killed and eight wounded.

"But though we did not suffer much from the enemy, we did a good deal from General Peck. This fidgetty old man kept fortifying and re-fortifying until his soldiers had become regular mud-diggers, and he had spent no end of labor and money in constructing works of immense magnitude, to defend a position not worth holding. There was digging and basket-weaving to an extent that went far toward developing the talents of the soldiers for farm work, and there were orders enough issued to supply the greatest army on earth. It will not easily be forgotten that the Eighth, who had been especially affected by gabion manufacture, awoke one morning and, instead of the stars and stripes, found a large sheet floating from their flagstaff with the inscription: "Peck's Avengers, or the Basket-Makers of the Nansemond."[2]

At four o'clock in the afternoon of April 11th, could be seen the pickets coming into town with a vengeance. Soon could be heard the long roll beating in the camps near General Peck's headquarters, and almost instantly the excited General himself came riding into camp at break-neck speed, the guard coming very near bayoneting his horse, ordering the regiment under arms immediately. Colonel Beach, who was in his tent, overhearing the order, came out and told the General "that he would frighten the best of troops, and that he (Beach) would not stir an inch until he received orders through the proper channels." As soon as the orders came properly

[2] Surgeon Mayor's Address. Re-union, 1867.

we fell in and marched to our position at the breastworks. Two days after, the rebels made an attack directly opposite the Sixteenth on the Somerton Road, but were so handsomely repulsed by the artillery, that they soon retired. From that time until the siege was raised we had the usual amount of hard labor and constant watching night and day that attends a siege and constant exposure to the enemy's fire. On April 24th, under the command of General Corcoran, the 13th Indiana, and the 11th and 16th Connecticut regiments went out on the Edenton Road on a reconnoissance. After skirmishing with the enemy for about thirty minutes, the regiment charged, driving the rebels from their pits to their earth-works, which was, perhaps, fifty rods. After holding this line long enough for the artillery to have a good duel and the General to find out the strength of the rebels, we returned within our defenses.

The regiment captured five prisoners, the officer of the pickets, a sword and various cooking utensils, which the rebels had left in their hurry. The casualties were one killed and seven wounded. This was a very successful skirmish and gave the men great confidence in themselves. Owing to swamps and the slashing on the edge of the woods, which the rebels had prepared, the men came into camp with their clothing completely ruined, making it necessary for an issue of clothing the next day.

On Sunday, May 3d, the regiment was ordered across the Nansemond river on the Providence Church Road, where they were engaged with the enemy several hours. The casualties were two killed and eight wounded. Privates H.W. Barber (A) and Frederick P. Cooley (H) were killed outright and Capt. Tennant, Serg't Pocket (D) and Corp'l Rivers (I) died from the effects of their wounds soon after, making really a loss of five killed. First Serg't Blakeslee (A) seriously wounded in the head, (making the second time in the same place,) was examined by Col. Beach, Capt. Pasco, and other members of the regiment and pronounced dead and left on the field.

Chaplain Francis B. Butler, of the 25th N.J. Regiment, while picking up Serg't Blakeslee, was fatally shot by a sharpshooter and died a few hours after with prayer on his lips for the wounded who lay around him. Under good surgical treatment by Surgeon Mayer, Serg't Blakeslee was able to fight other battles. Capt. Tennant was wounded in the early part of the action, and was taken from the field on a rude litter; notwithstanding the pain caused by the wound, he was cheerful and smiling; and remarked that he was good for a ten day's furlough. He was a brave young officer, and one of the best in the service. He was greatly beloved, and his early death brought sadness to many a brave heart. Young Barber's last words were "Tell mother that I never was a coward."

CHAPTER V.

1863.

PORTSMOUTH, ON TO RICHMOND.

After the siege was raised, the regiment remained in Suffolk until the middle of June, when they removed to Portsmouth, and encamped about three miles from the city, on the western branch of the Elizabeth river. This camp was formerly occupied by the 22d Georgia (rebel) regiment. The site was in a splendid grove and being on the bank of the river, afforded a fine place for the men to bathe, row, sail, and catch fish and oysters. I recollect of no place where the regiment lived so well, and enjoyed themselves so much as there. We had been there but two days when we were ordered to build a fort; but after working on it two days, it was abandoned by an order to be ready to march in light marching order with three days rations the next morning. Every thing was made ready and the men retired as usual at nine o'clock. But at half-past eleven the long roll sounded, and after forming in line the regiment marched through the woods to Portsmouth, a distance of three miles, in pitch darkness, and embarked on a transport, which left at 3 A.M.

YORKTOWN, WHITE HOUSE LANDING.

At eleven o'clock we found ourselves at Yorktown, encamping on the grounds where many a soldier had fought. The works built by McClellan and even those of the revolutionary war were still visible as was also the spot where the sword of Lord Cornwallis was surrendered by General O'Hara to General Lincoln, who was designated by Washington to receive it.

At half-past one on the morning of the 26th, the long roll beat again, and after falling in, the regiment marched to the wharf where they waited in the rain till 10 o'clock for a transport. At eleven all was ready, and after sailing up the Pamunkey river, which is one of the crookedest rivers in the country, as well as one of the finest and most picturesque, we disembarked at White House Landing.

Spier's cavalry had gone ahead of us one day and surprised the rebels, they barely having time to get aboard a train which stood in readiness to take them to Richmond. The cavalry then made a little raid through three counties, returning on the evening after the day of our arrival. They reported one severe fight and the capture of Brig. Gen. Wm. H.F. Lee, one Lieutenant-Colonel, one Surgeon, and one hundred prisoners. They also brought with them thirty-five wagons, besides burning eighty-five wagons and several railroad bridges. On the 28th, General Dix and Staff arrived and preparations were at once made for a move towards Richmond. Captain Pasco, of Company A, received his commission as Major on this day.

BLACKBERRY RAID.

On the morning of July 1st, the reveille sounded early and the division crossed the river, and after marching hard all day encamped for the night near King William's Court House. The next morning we were aroused at three o'clock, and marched to Brandywine, a distance of eight miles. On July 3d the reveille sounded at three o'clock, but we did not commence to march till five, when we marched pretty steady until one o'clock. The entire army had to stop then on account of the heat. This was the hottest day of the summer, and between the hours of twelve and one, sixty-five men fell out of the Sixteenth, fifteen of them having received a severe sunstroke. Four out of our Brigade died almost instantly. At five in the afternoon, the division commenced to march again and did not halt till nine in the evening, making in all about twenty miles that day. The name of this place was Taylor's Ferry.

The next day being July 4th, was a legal holiday in times of peace, but was not so for this division. The 11th and 16th Connecticut regiments and a section of battery were left at this point to guard a bridge and the supply train. The rest of the army moved to Hanover

Court House, where they had a rather tough time. After marching all day, they had to tear up three miles of railroad, cut telegraph wires and burn bridges, all under the fire of the enemy. This did a great deal of good, for Lee was in Pennsylvania, and finding his communication cut in the rear, retreated.

The next morning the Sixteenth was up at three, moved their position at five, and at eight the troops who had gone to Hanover Court House, returned. At noon three regiments were detailed to forage on the country and take all the beef, mules, horses, sheep and salt, which they could find. At two o'clock a supply train from White House Landing arrived with two days rations. At eight in the evening, we found that the enemy were after us, and we started on the retreat. Contrabands followed us in large numbers. At ten the rebels were close on our heels, as we passed through Mongoheck. At three o'clock the next morning we had gained somewhat on the rebels and were allowed an hour's rest at the Town of Ayletts. The men were so tired on this march, that they actually slept while marching, and when we halted at 3 o'clock that morning the men fell as if struck by lightning and were all sound asleep in two second's time. But our position was not safe, and the bugle sounded the reveille at four and after partaking of a scanty breakfast, we marched till half-past two in the afternoon, when we encamped about a mile this side of King William's Court House in a hard rain storm. This was certainly a hard fate for those who went to Hanover Court House, too fearful to be believed, making three days and two nights of hard labor with only one hour's rest. But it had to be done. Several who were unable to keep up and fell out were mercilessly shot dead by guerrillas who harassed the rear. A great many of the men provided themselves with horses, which they took out of barns by the road-side. The reveille as usual sounded at an early hour (three A.M.) the next morning and we marched to White House Landing, reaching there at eleven. The next day the division took up their line of march and marched to Hampton through a heavy rain storm and mud knee deep. Thence by boat they proceeded to Portsmouth, reaching their old camps on the afternoon of the 14th.

This raid in history is called the "Blackberry Raid," there being no end of blackberries, of the most luscious kind. It was during this raid that a new Chaplain, (Charles Dixon, of the Eighth Conn. Vols.,) reported for duty. There seemed to be at first serious objections, both among the officers and men, to our having a chaplain, but they were soon dismissed; for he was one of those rare men whom all learned to love, and his bravery in battle and southern prisons won him a good name. On our return to Portsmouth, through a special order, a Regimental Band was organized, which grew to be a most excellent one, and added not a little to our military standing.

CAMP TENNANT.

July 15th, 16th, and 17th, was spent in fixing up our camp. But the 19th, 20th, and 21st was spent entirely in trying to have a review. The first day we marched two miles, stood in a broiling sun for two hours, and then returned to camp. The next day the same performance was gone through with and no reviewing officer came. Finally, on the last day, just as General Foster got within a few rods of us the rain came down in torrents, and continued till General Foster went away. The rain stopped at five, and as the men were all ready for the review it was agreed that the ceremony should be gone through with by Colonel Burnham acting as reviewing officer, and Major Pasco taking the command. From this time until the ninth of September there was but little to attract attention, except the general routine of parades, picket and police duty. The regiment therefore had a little time of rest.

On the 30th of July, Colonel C.H. Prentice, J.G. Rathbun, J.A. Case, and J.S. Brooks, of Hartford, made a visit to the regiment. As nearly the entire regiment were acquainted with them, they were very welcome, and we presume they enjoyed their visit, especially the day that they visited the outposts, a distance of seven miles from camp, when one of those southern rain-storms came up, and they got a thorough soaking. This delayed their visit another day, in order to have their clothes dried. The quartermaster temporarily supplied them with brogans and clothes, but the amusing part of it was that Col. P. was so large and tall that the soldier's sizes could not be gotten only half on, and therefore with a cap and coat half on, and pants that reached a little below the knees, and with a huge pair of brogans, he looked so badly that he stayed in the Colonel's tent and played euchre all day and night. It was about this time that we had some terribly hot weather, the thermometer reaching 118° on August 2d, and 115° on August 3d. The pious members of the regiment went to work and built a chapel, which turned out to be a very fine edifice, and together with a temperance organization which started about this time, effected a great deal of good. The chapel was dedicated with impressive ceremonies, and a church organization was formed, the fruits of which last even to this day. The surgeons ordered some hospitals built, which were also dedicated by a fine dance. The morale of the regiment at this time was about perfect, and I can not better write of it than in the words of Surgeon Nathan Mayer, in his address to the regiment, at its reunion in 1867.

"Gradually the finest camp but one, which it has been my fortune to see, grew up. The most perfect order, the most civilized condition prevailed. The tents were neatly and prettily furnished, as our Connecticut country homes are, and the ground was always in beautiful condition. As winter approached the men built a hospital of logs, log-houses for the officers, log-kitchens, and eating saloons for the companies. Our pioneers erected a perfect village. Everybody purchased pine slats and made pretty huts, using their tents as peaked roofs to the structures. A chapel of considerable pretensions was raised—and here the difference between Connecticut and New York soldiers was apparent in more than one way, but in none more than in this. While the New York battery at our side thought first of all of erecting a theatre, we built a hospital and a chapel. While they gave their leisure to studying parts and rehearsing them, we organized prayer meetings, a choir, and endeavored in various ways to perfect ourselves as soldiers and men. Of great benefit was the presence of ladies in our camp. A number of officers and men had asked their wives to visit them in camp, and ere long a pretty row of cottages extended on the right flank of the grounds; in these there was singing, and chatting, and playing euchre or whist, and other social games almost every evening. And the demeanor of every soldier in the camp of the Sixteenth Connecticut at that time was gentlemanly. Not an oath was heard, not a phrase that in any way could offend the ear of a virtuous female; not an act perpetrated that savored of anything but deference and respect toward the sex. To complete the softening influence thus cast on the regiment the Colonel's mother, Mrs. Burnham, paid us a visit, and went among the men as such a good, sensible old lady, with a heart full of kindness, and a head full of shrewd Connecticut sense, would naturally do. She advised and comforted those in trouble, listened to complaints and nursed the sick. She was regarded as a representative of the good respectable home life which we had left, and to which we would sooner or later return, and the veneration and affection of the regiment for her was unbounded.

During all these weeks the military standing of the regiment rose perceptibly. There was not a cleaner, prompter, more loyal, reliable, and honest regiment in the service. No brighter arms, no quicker evolutions, no greater perfection in drill was to be found any where. The dress parade every evening gathered a crowd of lookers on. The guards, if detailed to other points, attracted attention. And the name of the Sixteenth was a good name in every man's mouth."

Excursion through Dismal Swamp.

On September 9th, five companies were detailed to go to South Mills, to do two weeks picketing. South Mills was about twenty-two miles from our outposts, and within seven miles of Elizabeth City. It was infested by guerrillas, and was merely held that no rebels might come through Dismal Swamp and surprise us. It was about the only place in that famous swamp where a few houses could be built and called a village. The detachment, under the command of Major Pasco, reached Deep Creek that morning, at half past nine. At Deep Creek the five companies, together with two wagons loaded with provisions, embarked on two canal boats, and hitching mules to the boats we were on the way to North Carolina. The scenery through the swamp was novel. This mode of traveling was new to the soldiers, and was highly enjoyed. Some footed it on the tow-path, some played chess, or cards, others smoked and read, and others danced to music from a fiddle and guitar, played by Dennison and Thompson. The canal was filled with turtles, snakes, and frogs of large dimensions. Thousands of snakes were killed that day, as the boat moved slowly along by the men cutting them in two with willow switches.

Arriving at South Mills at 7 P.M. we immediately proceeded to encamp in the dark by the side of Spier's cavalry about half a mile from the wharf; but before we could pitch our tents a terrible thunder storm came up and continued all night. Such a storm of lightning, thunder and rain, but few of us ever witnessed before. We were about drowned out by morning. The next day was a lively one. We changed camp in the morning, and three of the cavalry who were on picket were fired into by a party of about 40 guerrillas. One of the men was shot in three places and expired immediately. His horse was also shot in three places, the other two escaped all injuries. Two companies of cavalry were immediately sent out and scoured the woods, but could find nothing. One of our spies came in and informed us that the guerrilla camp numbering about 800, had found out our strength and position and they would probably attack us. So we moved our camp to our original position by the side of the cavalry for safety. In the evening a guerrilla, while trying to crawl through the lines, was caught by the pickets. The next day three men were arrested, one of which proved to be a guerrilla, and was sent to General Getty for sentence. His name was Finley. Late in the evening three more companies of the 5th Pennsylvania cavalry came to reinforce us. On September 14th a spy came into camp and informed us that the commander of the guerrilla band was in town, and that he was to leave that night with four recruits. Captain Robinson, with 26 picked men, were detailed to go out and lay in ambush, where he was to pass. But we lost them as they had passed there in the afternoon.

On Tuesday, September 15th, three scouting parties, consisting of the 5th and 11th Pennsylvania cavalry went out in the morning on three roads and returned in the evening, bringing with them five rebel soldiers and a colored man. On the 21st, five companies of the 15th Connecticut relieved us and the next day we returned to our old camp at Portsmouth. This trip, although dangerous on account of guerrillas, who were continually shooting at us if we strayed a short distance from our command, was much enjoyed by the men. The excursion we shall always look back upon with pleasure. The beautiful scenery, and many incidents and exciting times we shall never forget. The frightful noises of Dismal Swamp in the night cannot be imagined by one who has never been there. A night in Dismal Swamp is never forgotten.

After our return from South Mills there was another of what the boys called "a long spell" of quiet, the time being used up mostly by picket duty. Thanksgiving was observed as well as possible. On the 6th of December Sergeant Major Herbert Landon and 1st Sergeant B.F. Blakeslee received commissions as Second Lieutenants; and on New Year's Day Lieutenant Turner was promoted to be Captain. On the 3d of January, the Chapel was dedicated with ceremonies, the Chaplains of the 11th Pennsylvania, 13th New Haven, 5th Pennsylvania cavalry, and 8th and 16th Connecticut regiments taking part. The building was filled to overflowing and excellent music was furnished by a choir of ladies and gentlemen. In the evening a prayer meeting was held. The next day a few recruits came, being the only ones the regiment ever had, for while other regiments recruited largely, no one cared to enlist in the Sixteenth, a regiment noted for having severe hardships. And it was very unfortunate for those who joined us that day, for nearly all of them died in prison. On the 20th of January orders came to be ready to move at an early hour the next morning. In accordance with the order the men were up as early as one o'clock; and at precisely three o'clock the camp was fired and in a few moments nothing was left of the finest camp we ever had. Taking the cars we went to Portsmouth.

CHAPTER VI.
1864.
PLYMOUTH.

At Portsmouth the companies E, K, G, and B went aboard the steamer S.R. Spalding and the remainder on the Vidette. The weather was very fine and we had merry times and a fine sail around Cape Hatteras, reaching Morehead City on the morning of the 23d and proceeded thence by rail to New Berne. We left New Berne at midnight on the "John Farron" for Plymouth, and arrived there at midnight on the 24th. On the 26th Companies A, C, and H, under Capt. Hintz, went on a raid with other troops, under Col. Maxwell.

These companies returned on the 28th, and after that there were several raids into the interior, which the men enjoyed very much, as they had exciting times in breaking up rebel cavalry camps and capturing and burning up large quantities of cotton and tobacco, besides taking a number of prisoners. For some time the regiment lay here, going on raids, doing picket duty, and making such fine dress parades that it called forth the entire town every evening. One little incident which occurred here, and as it has been published, I will insert it as written by Serg't Maj. Robert H. Kellogg:

"There's one thing, at least, to be said in favor of Plymouth. It was the home of a few "true blue," loyal Southerners—a very few, however. They were hard to find, and I fear they are yet. The loyal men before spoken of, and some who were not loyal, were blessed with numerous daughters, fair to behold, but apt to have a few little weaknesses, such as 'dipping snuff' and smoking corn cob pipes. One of these men lived in a small house half way between the camp of the 16th and the western or left end of the town, and was blessed (or cursed, I doubt if he knew which at times,) with three daughters, and pretty ones they were. 'The prettiest girls I've seen yet!' was the emphatic declaration of each succeeding man who was lucky enough by dint of long watching or shrewd stratagem to get a peep at them. For, be it known, the father was as watchful over these fair scions of his house, as any ogre, read of in fairy tales, could possibly have been over his captives. Perhaps he had read some sensation tale of 'excesses of a brutal and licentious soldiery,' and thereupon resolved to keep his household uncontaminated from the least approach of such an insidious foe. I can not think he had taken a good square look into the honest faces of the 16th men, nor heard Chaplain Dixon preach to his crowded audience of boys in blue, every Sunday. At all events he seemed determined that no officer or soldier should form the acquaintance of his girls. On the other hand, our boys were quite as determined that they would become acquainted with them. But how was it to be done? That was the question which was presented to the mind of many a one who had cast 'sheep's eyes' at that humble dwelling in the hope of getting a glimpse at its fair inmates. Many and various were the plans which were made, but alas!

'The best laid schemes o' mice an' men,
Gang aft a-gley,
And lea'e us naught but grief and pain,
For promised joy.'

"None had been successful until at last one day two members of Co. "A" walked coolly and boldly into the forbidden cottage. First let me give the names of the ones who did it, then I'll tell how they did it. The persistent and successful schemers were Corporal Sam Belden, (remembered by every one of his surviving comrades to-day and by many friends in this vicinity,) and Private John Quinn. And this was 'the way the fort was taken.' After much polishing of buttons and brushing of uniforms, they obtained possession of the Company Clothing Book and another volume of similar size, which they found in the Orderly Sergeant's tent; and on a pleasant afternoon quietly left the camp, unnoticed, and proceeded to the scene of interest. A modest knock at the door brought out 'pater familias' or 'old tar heels' as the unsuccessful besiegers spitefully termed him. Corporal Sam coolly informed him, with that imperturbable gravity of countenance and manner for which he was celebrated, that they were deputed by General Wessells, who was in command of the Post, to take the census of the town. There was no getting around that, for an order emanating from such a source was not to be lightly disobeyed; so they were rather ungraciously admitted to the heretofore unvisited house—couldn't call it a mansion by any stretch of the imagination. Once seated inside, Corporal Sam as spokesman, commenced a series of questions which the U.S. Census Commissioners would have hard work to equal, private Q. jotting down the replies of the blushing and confused girls, and of the astonished father. Of course, by this cool and ingenious method they obtained the names of all, their ages, and other interesting information, and moreover they did it all with such suavity, and conducted themselves with such gentlemanly deportment, that, from that day they were invited, happy, envied, and regular visitors at the forbidden house."

On March 3d, at noon, the regiment was again on board the "John Farron," and on its way to Newbern, arriving there the next day evening and quartering in barracks near the Neuse River.

On the 11th the "Lancer" brought our baggage, and also the sad news of the drowning of Captain Mix, who in attempting to return to the regiment took a small schooner for Roanoke Island. While in Albemarle Sound, a squall struck the boat; and the boom knocked Capt. Mix overboard, as he sat smoking on deck. A boat was instantly lowered; but when within twenty-five feet, he sank. He was a good swimmer, but was encumbered with a heavy overcoat and large boots. He was well versed in tactics and military discipline, and was the last of the original captains of the Sixteenth.

On March 17th, Company "G," was ordered to Fort Stevenson, to relieve the Twenty-first, who were ordered to Little Washington.

On March 20th, a negro riot occurred across the Trent River. Captain Burke, with one hundred men of the Sixteenth, soon quelled it, bringing with him between two and three hundred prisoners, whom he turned over to the Provost Marshal.

Pack up at once, was the order soon after tattoo, and at midnight the regiment with all its baggage was aboard the "Thomas Collyer," returning to Plymouth. It was terribly stormy and rough; and at seven in the evening the vessel got out of the channel and ran aground in Albemarle Sound, a distance of about seven miles from Roanoke Island Landing. Here we lay until half past eleven on the morning of the 23d, the gale blowing terrifically, and the boat going higher and higher on the ground with every wave. The men were without rations, and suffering terribly from the cold and freezing spray. A flag of distress was raised, but not until the storm abated did any vessel dare come for our rescue. Finally the "General Berry," which was at Roanoke Island and had been watching for twenty-four hours, came and took us to Plymouth. The "Thomas Collyer" was nearly dashed to pieces, and it was some months before she was got off the bank, and was put in running order. The regiment lost considerable camp and garrison equipage, and some ordnance stores, which were washed overboard.

Battle of Plymouth, 1864.

I find in my diary, as early as March 24, that our pickets were fired into by rebel scouting parties, and on the next day we were expecting to be attacked. This rumor probably arose from some contrabands whom we traded with at the picket post, on the Columbia road, and who reported the enemy in large numbers in two counties south of us. These reports, together with the information General Wessells received, that the ram Albemarle was about completed, led the General on the 13th of April to ask for more troops, in order to hold the place if attacked. General Butler replied: "You will have to defend the district with your present force, and you will make such disposition of them as will in your judgment best subserve this end."

About the 14th of April, while officer of the picket, on the Lee's Mill road, an officer of General Wessells' staff and the officer of the day, invited me to accompany them outside of the lines, to see what information we could pick up. Mounting cavalry horses, we went out a distance of four or five miles, returning by the way of the Columbia road picket post. At one house where we stopped, a lady who had just arrived from the interior said that the rebels were concentrating, and it was reported that they were going to attack Plymouth. As I had heard these stories before, I paid but little attention to her report at the time.

On Saturday, April 16th, two days after, I was again officer of the picket on the Columbia road. The next morning (Sunday) at dawn, while asleep at the reserve post, I was awakened by the discharge of a musket by the picket at the bridge. Rushing to the spot, I found the picket to be William Maxwell, of Company A. He reported five or six scouts who had come to the edge of the woods suddenly, but fled on being fired at. I reported the fact to General Wessells, on being relieved at nine o'clock A.M. He seemed to think them guerrillas, but they proved to be advance guards, for in the afternoon when most of the soldiers were in church, the pickets were attacked by cavalry on the Washington and Lee's Mill roads simultaneously, and so sudden was the attack on the Washington road that the entire reserve picket were taken prisoners.

The "long roll" was sounded, and the troops prepared for the attack. Light artillery and cavalry were immediately sent out to ascertain the strength of the enemy. They had a short engagement, resulting in one killed, and Lieutenant Russell of New York Cavalry badly wounded. In the garrison, there were besides the Sixteenth Conn. Vols., the Eighty-fifth New York Volunteers, One Hundred and First, and One Hundred and Third Pennsylvania Volunteers, Twenty-fourth New York Independent Battery, two companies of the Second Massachusetts Heavy Artillery, two companies of New York Cavalry, and two companies Second North Carolina Volunteers, making in all 1,600 effective men. Early in the evening the enemy made a furious attack upon Fort Gray, on the river, a mile above the town. By eleven o'clock in the evening it was ascertained that the enemy had a force of between ten and twelve thousand men, and all loyal women and children in the place were embarked on board the "Massasoit," and sent to Roanoke Island. It was very evident to us that we must either be killed or go to

"Libby." Company "H," Captain Barnum, had been sent that morning to Roanoke Island for duty, and therefore a remnant of the regiment avoided the fate of prison life in the south.

The next day the enemy opened with artillery at an early hour, and the firing on the skirmish line was very lively until eleven o'clock. Captain Burke was wounded in the shoulder during the morning. At five o'clock in the afternoon I was detailed with fifty men to skirmish with the enemy on the Lee's Mill road for an hour or two to allow the regular picket line a little rest and time to eat. I had hardly got the line properly deployed, when it seems the enemy were ready to make their assault on the town. From the woods emerged the Confederates in great numbers. The loyal line fired a few regular shots, but the enemy came pouring out of the woods in such numbers that the Union line withered and shrank back. The enemy's artillery came to the crest of the hill, and so well was it manned that our camps were completely riddled, and Fort Williams partially silenced.

It was a regular artillery fight, and many old army officers said it was the handsomest artillery duel they ever witnessed. Three of the Sixteenth were wounded in the engagements in the skirmish line, one of whom was A.P. Forbes, of Company B. The enemy came on so rapidly, and we retired so slowly, that the two lines nearly met. One of the Sixteenth was pressed so closely that, in the dusk of the evening, he dodged behind a stump and thereby saved himself from capture. He was so near the Confederate battery that he overheard a staff officer give the order, "It is no use, captain, we cannot endure this fire,—limber to the rear." The enemy retiring, he returned inside our ranks.

Heavy artillery firing was kept up until eleven o'clock P.M., and under cover of the darkness, the enemy advanced up to Fort Wessells, a work about ten hundred yards in front of the line of fortifications. Fort Wessells was furiously stormed three separate times, by a very superior force with great loss of life. The third time it had to succumb, and sixty men were captured. The fort was well supplied with hand grenades, which were used with great effectiveness. It was during this night that the famous ram, "Albemarle," came down the Roanoke river, passing our batteries, sank the Southfield, and drove off the balance of the fleet of gunboats. The Bombshell had previously gone up the river, and in returning was so completely riddled by the enemy's batteries, that she sank on arriving at the dock.

The next day their artillery opened on the right of the town, but the lay of the ground in front of our works was such that it was really our strongest point. A few of their infantry advanced into a ravine in our front, and were unable to extricate themselves until dark.

During the day the town was pretty effectually shelled, and a caisson was blown up on each side. Our navy being entirely gone, the ram "Albemarle" did good service for the enemy, with ninety and two hundred pound shot.

Three separate times were we asked to surrender and save further sacrifice of life, but each was peremptorily declined. General Hoke (rebel) the last time replying, "I will fill your citadel full of iron; I will compel your surrender, if I have to fight to the last man."

There was no doubt now but what we must succumb sooner or later. There was no hope unless reinforced, and that could not be as long as the ram was in the river. The men built bombproofs and traverses, which were a great protection.

Late in the evening, Co. "G" was ordered to the left of the town, on the Columbia road. They lay there during the night, preventing the gunners on the ram from sighting their guns and coming on deck; they also had two little brushes with cavalry, who broke through the line to procure beef that was in a yard near by. The enemy, meanwhile, were concentrating nearly half their force opposite this point.

By 4.30 o'clock on the morning of the 20th, (I find from their accounts,) they had ten regiments of infantry, four battalions of artillery, (Pegram's, Blunt's, Marshall's, and Lee's,) and two companies of cavalry, besides the "Albemarle" and "Cotton Plant." This must have made a force of five or six thousand in line about six hundred yards in front of our works. At this hour a rocket was sent up as the signal for the attack, and a more furious charge we never witnessed. Instantly over our heads came a peal of thunder from the ram. Up rose a curling wreath of smoke—the batteries had opened, and quickly flashed fierce forks of flame—loud and earth-shaking roars in quick succession. Lines of men came forth from the woods—the battle had begun.

Company G, being on the skirmish line, fell back and entered "Coneby redoubt," properly barred the gate and manned the works. The enemy, with yells, charged on the works, in heavy column, jumped into the ditch, climbed the parapet, and, with the artillery company (who had previously occupied the redoubt), for fifteen murderous minutes, were shot down like mown grass. The conflict was bloody, short, and decisive. The enemy were in such numbers that we had to yield. The gate had been crushed down by rebel shot, and the enemy poured in to the number of five or six hundred, with thousands on the outside. Great confusion then ensued; guns were spiked, musket barrels bent, and all sorts of mischief practiced by the Union soldiers, while the enemy were swearing at a terrible rate, because we would not take off equipments and inform them if the guns could be turned on the town, and in trying to reorganize their troops, who were badly mixed, to take the next work. We were prisoners, and as we marched out of the fort we could see at what a fearful cost it was to them. There were in the fort at the time, forty artillery men, who fired grape and canister, and forty-two of Company "G," (two being unable to get in, or not hearing the orders, went back to town,) making a total of eighty-two men, against five or six thousand. Our loss was one wounded, an artillery man, while the rebel loss, from their latest accounts, was five hundred killed and wounded. The enemy then passed in the rear and on the bank of the river, to the right of the town, and while part of their force was on the right working towards the center, those on the left were doing the same. Every position was obstinately maintained. A squad of men here, and a squad there, the redoubts and forts were but slowly captured. For three or four hours, Fort Williams, with guns turned, did murderous execution, nearly two hours of which was in the streets of Plymouth. By half-past ten o'clock the last gun had been fired, the flag over our citadel lowered, and Plymouth had fallen.

The troops were captured by an overwhelming force, after one of the severest fights of the war. In the words of J.W. Merrill, the author of "Records of the Twenty-fourth N.Y. Battery," "there is no question that the defense of Plymouth by its garrison of 1,600 men against a besieging force of 12,000 men, was one of the hardest fought battles of the war." The rebels raised the "black flag" against the negroes found in uniform, and mercilessly shot them down.

The shooting in cold blood of three or four hundred negroes and two companies of North Carolina troops who had joined our army, and even murdering peaceable citizens (as I have the personal knowledge of the killing, with the butt-end of a musket, of Mr. Spruell, the man whom I boarded with, and by the way, a secessionist, for objecting to the plundering of a trunk which he had packed), were scenes of which the Confederates make no mention, except the hanging of one person, but of which many of us were eye-witnesses, was but the Fort Pillow massacre re-enacted.

The following order was issued the day after the capture by Gen. Peck:

Headquarters Army and District of North Carolina,

Newbern, N.C., April 21, 1864.

General Orders, }

No. 66. }

With feelings of the deepest sorrow the commanding general announces the fall of Plymouth, N.C., and the capture of its gallant commander, Brigadier-General H.W. Wessels and his command. This result, however, was not obtained until after the most gallant and determined resistance had been made. Five times the enemy stormed the lines of the General, and as many times were they handsomely repulsed with great slaughter, and but for the powerful assistance of the Rebel iron clad ram and the floating sharp-shooter battery, the Cotton Plant, Plymouth would still have been in our hands. For their noble defense the gallant General Wessells and his brave band have and deserve the warmest thanks of the whole country, while all will sympathize with them in their misfortune. To officers and men of the navy the Commanding General tenders his thanks for their hearty co-operation with the army, and the bravery, determination and courage that marked their part of the unequal contest. With sorrow he records the death of the noble sailor and gallant patriot, Lieutenant Commander C.W. Flusser, U.S.N., who, in the heat of battle, fell dead on the deck of his ship, with the lanyard of his gun in his hand.

The Commanding General believes that these misfortunes will tend not to discourage but to nerve the Army of North Carolina to equal deeds of bravery and gallantry hereafter. Until further order, the headquarters of the sub-district of the Albemarle will be at Roanoke Island.

The command devolves upon Colonel D.W. Wardrop, of the Ninety-ninth New York Infantry.

By command of

Major-General G. PECK.

J.A. Judson, Assistant Adjutant-General.

While I have given an extended account of the position and service, of Company "G," the remaining companies were not less exposed, nor did less fighting. The losses in the Sixteenth were one killed and twelve wounded. The total losses in the garrison were fifteen killed and about one hundred wounded. The number captured was about sixteen hundred. The accounts in the northern papers were meagre and greatly exaggerated, giving generally the losses of killed and wounded equal to the entire garrison, and the losses of the enemy in equal proportion to their number. The Richmond papers gave long and very accurate accounts of each day's operations, but suppressed their losses and the killing in cold blood of the North Carolina soldiers, negroes, and citizens. Their latest accounts, however, admit from four to six hundred killed and from fifteen hundred to two thousand wounded. The Rebel Government considered Plymouth of great importance, and promised Gen. Hoke a Major-General's commission if he would take the place. It was this incentive that made him storm the works with such desperation. From the Richmond papers we find mentioned as operating against us sixteen regiments of infantry, eleven batteries divided into two battalions and commanded by Majors Reid and Mosely. The artillery, together with two companies of cavalry, were commanded by Col. Deering. In addition to the above were the ram Albemarle and consort Cotton Plant. The whole force was divided into three brigades, (Hoke's, Ransom's, and Kemper's) commanded respectively by Col. Mercer, Gen. Ransom, and Col. Terry.

EXTRACTS FROM RICHMOND PAPERS.

The comments and correspondence in regard to the capture of Plymouth contained in the Richmond Whig, Dispatch, Enquirer, and Examiner, would fill three volumes of the size of this. Only a few of them can be inserted:

From the *Enquirer*, April 22, 1864.

CAPTURE OF PLYMOUTH—ONE THOUSAND SIX HUNDRED PRISONERS AND TWENTY-FIVE PIECES OF ARTILLERY CAPTURED.

The following is a copy of a dispatch received in Richmond, yesterday morning, by General Bragg:

"Plymouth, April 20th.—To General Bragg.—I have stormed and carried this place, capturing one Brigadier, one thousand six hundred men, a quantity of stores, and twenty-five pieces of artillery.

R.D. Hoke, Brig-Gen.

From the *Enquirer*, April 25th:

Promoted.—Brigadier-General R.F. Hoke has been promoted to the rank of major-general, to date from the capture of Plymouth, N.C.

From the *Enquirer*, April 26th:

The Plymouth Affair.—The Wilmington *Journal,* of Saturday, says "our loss in killed and wounded is not large considering the magnitude of the enterprise; but, as might have been looked for from the character of the conflict, the works having been stormed, a large proportion of the wounds are of a desperate character." When a place is taken by storm, and there is resistance, as in this case, the fighting is done hand to hand—guns are fired off at a trifling distance and the wounds inflicted in most cases are serious if not mortal. We learn that some of our wounded who have been brought to Wilson, bear evidence of the desperate character of the struggle whilst it lasted. They are wounded in almost every imaginable way, and but few of their hurts can be called slight.

From the Dispatch, May 2d, 1864.

A correspondent of the Raleigh (N.C.) *Confederate*, sends that paper a history of the capture of Plymouth, which is very interesting.

Feint on Warren Neck.—On the night of the 17th, an attack was made upon Warren Neck, under the direction of Colonel Deering. A gunboat of the enemy coming to the assistance of the garrison was sunk, and a force of infantry sent from the town was repulsed; but the enemy successfully resisted all attempts to take this stronghold. On Monday our artillery opened vigorously on the town; and during the day both parties pounded away at each other incessantly; but beyond a little skirmishing with the enemy and manœuvering for position our infantry did nothing. Toward evening, however, it became evident that something was on foot; and Ransom's brigade, and the 8th N.C. was drawn up in the woods facing the works on the Washington, Lee's Mill and Bath roads. A heavy line of skirmishers was thrown out under the command of Captain John Pegram, A.A.G., and advancing rapidly with the peculiar gait of sharp shooters and the yell with which our boys go to the charge, drove the enemy back into his works, and approached within two hundred and fifty yards of the fort, earnestly demanding to be led into the place. Meanwhile Pegram's battery dashed forward at a run, supported by the infantry, and unlimbering, devoted a furious fire upon the place. Three times we advanced, each time nearer, until within good charging distance; but the artillery had it all to themselves. The movement was merely a demonstration to call off the enemy's attention from Hoke's attack upon Fort Wessells, which, after a sharp but short resistance, fell before the superior powers of Hoke's brigade; and that night's work was done.

The Gunboat appears on the scene. Leaving a line of pickets on the field, the main body of the troops withdrew to prepare, by a few hours rest, for the attack, which we all felt would be made on the morrow, and as we lay down by our fires, every one wondered at the Albemarle's delay, and prayed for her speedy arrival. At three in the morning we were all awakened by the thunder of her Blakely guns, as she defiantly saluted Warren Neck, en passant, and sailed safely by over the obstructions which the enemy had placed in the river. She went to work at once among the enemy's gunboats, sinking one and driving the rest to Hatteras, and then turning her attention to the fortifications, she kept up a speedy fire during the morning, silencing the enemy's guns, and driving him into his bombproofs. But still the "stars and stripes" floated over his works, and as he refused, when summoned to strike his colors, it became necessary for us to do it ourselves, and the evening and night of the 19th were devoted to preparing for the assault on the morning of the 20th. Kemper's brigade had fought gallantly at Warren Neck. Hoke's men had taken Fort Wessells with three guns and sixty prisoners. It was now Ransom's turn.

The Columbia road which enters the town at its eastern extremity, running parallel with the river and near it, crosses Coneby creek about a mile from town: To this point Ransom's brigade, the 8th N.C. and Pegram's battery marched late in the evening of the 19th, behind a screen of woods, which hid the movement from the enemy, and reached the creek about sunset. The bridge was destroyed, and the creek was too deep to be forded. A strong picket of the enemy was on the opposite bank behind entrenchments, and about three-quarters of a mile off were two 32-pounders and five 12-pounders bearing on the spot. The pontoons must be laid for the infantry and artillery to cross, and that quickly or the movement would be a failure. The moon was shining brightly, turning night almost into day, and not a breath of air was stirring, so that every movement we made could be distinctly heard or seen by the enemy. Lieutenant Marshall Lee, with twenty men of Co. E 24th, was advanced to the water's edge supported by the rest of the company, and Co. A of the 35th, the whole under the command of Captain Barna Lane. The pontoon train, under Lieutenant Pool, 10th N.C.T., dashed down at a gallop, slid one boat into the creek and quickly and rapidly the two companies crossed and were immediately engaged with the enemy. The 24th followed at once, the men coming into line as fast as they got over, and the enemy fell back, closely pursued by Captain Lane's command, deployed as skirmishers. The pontoons were now laid and by 8 o'clock the infantry was over and formed in line of battle, the left resting on the road and the right on the river in the following order: First on the right the 55th, second the 25th, third the 8th, fourth the 36th, fifth the 24th, with two companies on the south of the road. In this position, we lay during the night, sheltering ourselves as well as we could from the enemy's two works on the road, which kept up a constant fire until nearly day break. Just before day a strong line of skirmishers was thrown out before the brigade under command of Captain Durham, Q.M., 29 N.C.T., but acting temporarily on General Ransom's staff. The Assault. At just 3-½ the morning of the 20th our line began to move forward, slowly at first, dressing on the center, and halting occasionally for that purpose. From the start the fire from the enemy's batteries was rapid and severe, striking down many a brave fellow; but closing up the gaps, the long line moved silently on, the left still resting on the road, till Pegram's battery, dashing forward at a run unlimbered in front, and opened fire on the enemy's works.

Then for the first time that morning our boys gave a loud yelling of defiance, and quickening their pace to a double-quick, pressed with a determination not to be resisted right upon the enemy's two works, which were taken with scarcely a moment's delay, the one on the south of the road, by the left of the 24th, led by Colonel Clarke and the one on the right by the right of the 24th, assisted by the 25th. The enemy fled in terror to the houses, Fort Williams, and any other place which suffered them protection from the fierce fire of our pursuing ranks. We were now in the town, and the head of every street running east and west was held by one or more of our regiments; but their position in line was somewhat changed. The 24th was still on the Columbia road, now street, with the 56th and 25th to the right, and the 35th and 8th to the left. Halting a moment to breathe the men and dress the lines, we pushed slowly and carefully forward, clearing the enemy from every street, yard, and house, from the windows of which and from behind the fences they poured an incessant fire. But nothing could check our progress and in an hour the enemy were all driven into Fort Williams or the entrenched camp. The fort was on our left and the camp in front. Leaving the 35th, the 8th, and a portion of the 24th, to contend with the fort, the rest pushed on for the camp, which the 24th being on the direct road soon reached and opened fire, exposed still to a severe musketry fire from the fort on the flank and the camp in front. In a few minutes the 56th came up on the right by another street, and by their arrival decided the contest, for immediately on the appearance of this additional force the enemy threw down his arms, and raised the white flag. Captain Lockheart, of the 56th, ran in to receive the surrender, and instantly both regiments poured into the camp, and throwing down their own foul guns and empty cartridge boxes, took the clean, well-filled ones, which were lying about, and pressed on through the tents to the western side of the camp, where they could see the United States flag floating over Fort Williams, evidence that the fighting was not yet done. Here we were shortly joined by the 43d North Carolina of Hoke's brigade, who came in from the west, having been delayed by a morass, which they had great difficulty in crossing.

The Town in our own hands. The town was ours. Every house and street in it was swept of the enemy, who shut up in his stronghold, still refused to surrender. This was a case for the artillerist and the guns of the captured forts were soon turned upon their stronger brother, fighting as the result shows, better for us than they had against us. Captain Cooke too of the Albemarle dropped some of his 90 pound shell among them. Still they resisted stoutly, showing a disposition to die rather than to yield. At last, however, some of our boys creeping forward through the entrenchments, got an enfilading fire upon them, which soon brought them to terms, and hundreds of them rushed out of the fort without arms and surrendered. Just at this moment a shell burst directly on the magazine, and when the smoke cleared away the hated flag was fluttering rapidly down to the ground. Without waiting for orders, the brigade swarmed into the fort, Company B, of the 24th, leading, and the color of everything was quickly changed from blue to grey."

From the *Examiner*, April 22d:

"The details of the affair at Plymouth are not yet fully ascertained. It is safe, however, to pronounce it one of the most brilliant affairs of the war. Its immediate fruits are important. Those which may ultimately flow from it, the possession of an abundant country, and the possible command of the inland waters of North Carolina, though as yet the subject of uncertain speculation, may well excite high hopes for the future. General Hoke, judging from the large number of his prisoners, does not seem to have made such thorough work as that by which Forrest has so shocked the tender souls and frozen the warm blood of the Yankees. The resistance he encountered was probably not as desperate, and the blood of the victors not so heated; though in a fortification carried by storm the loss of the garrison must inevitably be large. The strict laws of civilized warfare acknowledge the power of the victors to put all to the sword in such cases. However severe such an example might seem it would strike a salutary terror into the Yankees, which would be useful to them in the end, and their melancholy whine, at meeting a part of the punishment their barbarities merit, is absurd."

From the *Examiner*, April 25th:

"Tuesday evening a flag of truce was sent to Fort Williams, demanding the surrender of the enemy. The flag was taken by Colonel Deering and another officer; and General Wessels, the Yankee commander, refused to treat with them, but requested a conference with the General commanding. At the interview which ensued the Yankee commander said to General Hoke, that if he surrendered he would be sacrificed by his Government, and, he feared, would be retired from the service. "Then," replied General Hoke, "I understand that you are fighting for your commission and for no other cause. If such is your reply, I have only to compel your surrender, which I will do if I have to fight to the last man." The general assault followed Wednesday morning. It was made by all our forces. As our troops came within range of the enemy's artillery, they suffered very severely, as the ground in front had been surveyed and was staked off with target posts for artillery practice. Latham's battery had been placed just by one of the targets, and was shelled with such skill by the enemy that all his horses had been killed. The accounts of his casualties are deplorable, and we trust they may be reduced by the more exact statements, which will be officially given."

From the *Examiner*, April 30th:

"The Capture of Plymouth in the North. The news of the fall of Plymouth had reached the North. Of course, after their several days of felicitation that "the fort would surely hold out," this news was sudden and unwelcome to them. The Tribune announces it under this imposing head, in very large capitals: Surrender of Plymouth. General Wessels and one thousand five hundred men prisoners. Our loss one hundred and fifty killed. The rebel loss one thousand and seven hundred killed. North Carolina troops taken out and shot after surrendering. All negroes in uniform also murdered.

It is positively affirmed that the rebels in taking possession of Plymouth, ordered out the North Carolina (Union) troops, who formed part of the garrison, and shot them; and that all negroes found in uniform were murdered. We presume the account is correct, and it only proves that what was supposed to be an exceptional barbarity at Fort Pillow, has been adopted as the deliberate policy of the rebels. As the issue is to be made it must be met."

From the *Examiner*, May 3d:

(Extract from a letter on the *Victory of Plymouth*, dated Plymouth, N.C., April 24, 1864.)

"During Monday night Hoke's and Kemper's brigades slept on their arms in the position they had gained. Before day break next morning three regiments of Ransom's brigade and Col. Branch's artillery were ordered to support them, and Ransom, with two regiments and artillery, was again ordered to the right to make a demonstration. At light the enemy opened a heavy artillery fire upon our position, to which we replied, also turning their own guns from the captured fort upon them. General Hoke, after making a more thorough reconnoissance on Tuesday morning, did not attack on the left, hence returned Ransom's regiments to him on the right.

About midday he determined to send Ransom's brigade, with artillery, to the right, Coneby's bridge a distance of four or five miles, to make a simultaneous demonstration with him, while he would attack from his position on the left with his and Kemper's brigades. Ransom reached the bridge about dark, threw forward his skirmishers, who found the enemy in strong position on the opposite side, and the bridge destroyed. Finding the enemy were in the rifle pits and stubbornly refusing to yield the position, three pieces of artillery, under. Captain Blount, were advanced to within three hundred yards of the bridge and the enemy were soon dislodged. Our sharpshooters again advanced and the enemy re-appeared when some gallant fellow of the 24th N.C. regiment plunged into the creek, brought back a skiff, and immediately a party of select men were put over in it, and ordered, at all hazards to drive the enemy. The pontoons were hurried to the front, one placed in the creek, which ferried over three or four companies that followed, deployed as skirmishers, and the enemy fled, abandoning a position of vital importance to them. The pontoons were soon laid, and the infantry consisting of the 24th N.C., Col. Clark; 25th Alabama, Col. Rutledge; 56th N.C., Col. Faison; 35th N.C., Col. Jones; and (of Clingman's brigade) the 8th N.C., Col. Murchison, passed rapidly over and deployed into line at about a mile from the enemy's outposts, the right flank resting on the Roanoke, and the left on Coneby Creek. The artillery commanded by Col. Branch, did not cross, as the enemy could easily hear the crossing, and would shell furiously at the least noise. About twelve o'clock the troops were all in position, and were ordered to sleep on their arms and rest for the heavy work just ahead of them. The night was perfectly calm and cloudless, with the full moon shedding its soothing beams upon the sleeping veterans as they lay upon the bare ground, covered with their blankets in groups of two or three for warmth, as the air was sharp and piercing, seeming not even to dream of the morning's carnage. The field officers in rear of their regiments paced to and fro unable to rest from the heavy responsibilities that were crowding upon them. Along the line of the skirmishers commanded by "the fighting Quartermaster Durham" and the gallant Applewhite, the quick blaze of the rifle, like fire-flies at night, was the only relief from the dead calm that prevailed around, as the enemy's heavy guns, however, were belching forth shell and spherical case, firing with great accuracy. When Generals Hoke and

Ransom separated, it was understood that as soon as the latter was in position, he would signalize the fact by a rocket, when General Hoke, with his and Kemper's brigades, would attack on the left, and Ransom on the right, would make an attack or a demonstration, as he thought best. After making a reconnoissance, Ransom at one o'clock in the morning (Wednesday), dispatched General Hoke that he was in position, but would defer all movements until the dawn of day, when he would not demonstrate, but attack, and intended to carry the place by assault, asking from General Hoke his simultaneous co-operation. He called together his field officers, communicated to them his purpose and plans, and by his confidence, coolness, and resource seemed to inspire them thoroughly with his own self-reliance. As a ruse, he determined to place his artillery in the rear of his infantry, and thereby cause the enemy to overshoot our lines, which proved a complete success. The moon had just gone down and the gray streaks of the morning were faintly visible in the east, when the signal rocket went up, and the line of skirmishers were ordered to advance, which they did handsomely, driving in the enemy easily. The infantry now moved forward, and the artillery, consisting of Blount's, Marshall's, and Lee's batteries, under Colonel Branch, dashed forward at a full gallop into position, and opened immediately upon the town and forts at about twelve hundred yards. The enemy had by this time concentrated a most terrific fire from their siege guns and field pieces. Just at this time General Hoke opened, with his artillery under Majors Mosely and Reid, a very rapid and tremendous fire, and his infantry sent up yell after yell as if charging. Ransom caught up the sound, and rising in his stirrups, from the head and right of the line, in a clear and ringing voice, gave the command, "Charge, boys, and the place is yours," and such a charge and yell no one who ever saw or heard it can ever forget. It baffles description. It was as the wildest gust of the tornado as it prostrates the forest, or the mad fire as it dashes through the prairies, it was a thunder bolt. The hail of shot, shell, canister, grape, and Minnie balls were not only unheeded but seemed unheard. The infantry and artillery fell thick and fast, but immediately the gaps were closed. In ten minutes the two outer forts, with eight guns, were captured, our infantry scaling their parapets, and the infantry within one hundred and fifty yards of the forts, horses, and limbers blown up and cannoneers shot down, and yet those remaining stood to their guns without shelter, confident of victory, and to avenge their dead. The whole command of officers and men, infantry and artillery, seemed enthused with the inspiration of certain victory. Several hundred prisoners were captured in these forts, which were immediately sent to the rear, and now began the contest for the town more than a half a mile in length, the enemy's infantry slowly retiring and stubbornly resisting our advance, Fort Williams dealing out grape and spherical case; their field pieces at the farther extremities of the broad straight streets raking them with a murderous fire; their infantry in the houses and cellars and behind fences, delivering galling charges of Minnie shot, but all of no avail. Our men were confident, aroused, and irresistible. They pressed on steadily, without halt or hesitation, tearing down fences, hedges, and every obstacle that they met, capturing the enemy at every step. Soon they were in a rout, and our infantry pressed through the town and waved their battle flags over the fortifications in front to General Hoke's forces, who could scarcely believe that so much could be accomplished in so short a time. * * Ransom's charge has not been surpassed at any time; his military genius comprehended the situation, and he was master of it. He determined upon his charge, knowing what pluck and dash could accomplish, and satisfied it was the only point of attack. With twenty-two hundred muskets and nine pieces of artillery, he charged the enemy's work which they regarded unassailable and carried them without a halt. His loss, which reached five hundred in thirty minutes, shows how sanguinary was the work." (Signed R.)

From the Richmond *Dispatch*, April 30th:

The Fall of Plymouth. A Specimen of Yankee Lying. A Sympathetic order for General Wessels. Negro Soldiers Butchered. "The Philadelphia Inquirer contains the official and other announcements of the fall of Plymouth. The butchery of the negro troops is news here, though if General Hoke had butchered the whole garrison in the assault, after a refusal to surrender, it would have been perfectly proper under the laws of war. It will be seen that the loss of the Confederates is put down at fifteen hundred!! The following is a telegram dated Fortress Monroe, the 24th instant: The gallant garrison at Plymouth, after a desperate struggle with the rebel foe, who besieged them by land and water, with an infinitely superior force, were compelled to surrender, but not until they had slaughtered hundreds of the enemy in their attempts to storm the forts. The fight commenced late on Sunday afternoon, and continued until half-past ten on Wednesday morning, when the surrender was made. Our men fought with the ferocity of tigers, and they would never have yielded had there been anything like an equality of forces. But the rebels outnumbered them nearly ten to one. When the attack first was made the Confederates were twelve thousand strong, and afterwards received eight thousand more as reinforcements. Aided as they were by their powerful ram and gunboats, it is not at all surprising that they succeeded in capturing the Town of Plymouth."

From the New York *Herald*, April 26th, 1864:

"The Rebel Losses are, beyond the slightest doubt, immensely heavy. When it is considered that every fort around Plymouth was stormed from three to seven times, and each assault repulsed with great slaughter, besides pouring broadside after broadside into the rebel ranks from the Miami and Southfield, the casualties among the rebel troops must have been enormous. A rebel surgeon was heard to say that "the damned Yankees had killed and wounded one-third of their whole force, and he hoped that no mercy would be shown the cursed Yankees." The gunboat Whitehead went on a reconnoissance, on Wednesday, a short distance above Plymouth, and the officers and crew observed about three hundred rebel troops engaged in burying the dead. From a steeple on the town church, overlooking a large tract of land, it was found that the field of Asa Johnson (about sixty acres), was completely filled with dead and dying rebels. The entire rebel force could not have been short of from fifteen to twenty thousand men, of whom one-third are unfit for future service.

The Fort Pillow Massacre Re-enacted. My informant also acquainted me with the fact that all the negroes found after the surrender, were stripped of their clothing and brutally murdered in cold blood. It must be understood that General Wessels had no colored troops at Plymouth, save a few recruits for North Carolina regiments, and the poor unfortunate blacks thus butchered were merely laborers for the government. The negroes were formed into line, in a nude state, and fired at by the brutal soldiery, purporting to represent Southern chivalry. Nature revolts at these facts; and the plan apparently adopted by the Rebels for the future disposition of the negroes is emancipation from, and not for, life."

CHAPTER VII.
1864.
PRISON LIFE OF THE OFFICERS.

I shall not in this chapter pretend to give a complete history of Prison Life in the South, only a part. As I look back over the great mass of events that were continually occurring during all hours of the day and night of less than one year, I give up in despair. The inhuman treatment, and fearful atrocities of our brutal keepers are heart-rending, and I can poorly illustrate them. Language is unable to describe the real condition of affairs in southern prisons. No one can present in its true light the fearful suffering experienced in them. Others have contributed sketches for the dark picture. I shall therefore be very brief.

By noon of the 20th of April, the entire garrison of Plymouth was in the hands of the rebels, and all the captured were concentrated on the Washington road. During the afternoon, the rebel cavalry scoured the woods and shot dead every colored soldier and man that was able to bear arms. The number murdered in this way must have been in the neighborhood of one hundred.

We went to sleep that night feeling very gloomy, tired, and hungry, most of us having had nothing to eat during the day. It was nothing new to us to sleep on the ground without anything over us, for we were quite used to that, but we could easily see what we were to expect from the hands of the rebels, and many were thinking of home and how anxious our friends would be to hear of our fate, whether killed in the battle or captured; and if captured, worse than killed. The next day we had twenty-five hard crackers and on an average about ten ounces of raw salt pork issued to us. This had to last us till the morning of the 26th, making five days, thus making five crackers and two ounces of pork per day. On this we had to live and during the day make quite long marches.

Colonel Beach, Lieutenant Colonel Burnham, Major Pasco, Quartermaster Robins, Surgeon Mayer, Assistant Surgeon Nickerson, and Captain Burke, together with the wounded men, remained at Plymouth. About noon of the 21st we took up our line of march, halting about nine in the evening, making a distance of thirteen miles. We marched more or less every day until the noon of the 25th, when we arrived at Tarboro, having passed through the villages of Foster's Mill, Jamesville, Williamston, and Hamilton. Crowds of women and children lined the roadside eager to get a glimpse of the "Yankees," some hissing and insulting us in various ways. On the 23d, the 11th Virginia guarded us, and we learned that the same regiment fought against us on the Edenton road at Suffolk a year before. By the time we reached Tarboro, we were nearly starved, having been out of rations from twelve to thirty-six hours. At noon, some bacon and corn-bread were issued to the officers, and to the enlisted men in addition some peas which were by long cooking softened sufficiently to be eaten. At one o'clock, April 26th, all the officers and the 85th New York went aboard the cars and were on their way to the famous prison pens of the South, thus separating the officers from the enlisted men—only a few of whom lived to meet again at New Berne a year after. Traveling all night in crowded cars, we arrived at Wilmington at daybreak, where we crossed the river by ferry boats, and again taking cars we arrived at Florence, South Carolina, at 10 P.M.

Here we again changed trains, being packed by fifties in box cars. This was a miserable night, as we were packed so closely that we could neither lie down nor sit. Some had to stand up, and those farthest from the door nearly suffocated. Leaving Florence in the morning we arrived at Charleston at 10 P.M. Here another change of cars was made, sixty being packed in each car, and we remained there all night. The next morning some hard-tack and good bacon were issued. Leaving Charleston at seven in the morning we arrived at Savannah at two P.M. Here more rations were issued and another change of cars was made. We finally arrived at Andersonville at 2-½ o'clock the next afternoon. Having marched us to a position, where we could obtain a view of the famous pen, Capt. Wirz soon appeared on a horse, making such a comical appearance that we laughed outright. "You Got damn Yanks, you vont laugh when you get in dat Bull Pen," were his first words of greeting. Having been sent to Andersonville by mistake we were not allowed to go in to the Bull Pen, but were marched to a school house or church, where we staid until the next day. Being Sunday, Chaplain Dixon gave us a short sermon from Psalms 37th, which was very appropriate and suitable under the peculiar circumstances in which we were then placed. At half past nine in the forenoon we were placed aboard the cars again, and went to Macon.

CAMP OGLETHORPE, MACON.

Here we went into Camp Oglethorpe, one hundred and ten in number, being all the officers captured at Plymouth, and going under the title of "Plymouth Pilgrims." General Howell Cobb soon rode up on his charger with a bran new saddle and bridle, and introduced himself by saying, "that we were just the fellows he wanted to see." "Why don't you go to the front then," replied Adjutant Clapp. This rather astounded the old gentleman. At this camp one day, Adjutant Clapp was having his haircut and accidently the barber nicked his ear a little. A lady who was at the time passing around the camp noticed it, and the next day sent a little brother into the camp to sell pies, and in the bottom of the basket was a book with a note directed to the soldier with the cut ear, and stating that she was from the north and sympathized with our misfortune. For good reasons I withhold her name, but she was a noble and brave girl, and afterwards risked her life in aiding Federal officers to escape.

On the 17th of May, 900 officers from Libby arrived and were placed in a pen which had been built a short distance from our camp, and in the evening we were also placed there. Here we found Lieutenant-Colonel Burnham, Major Pasco, Quartermaster Robins, and Captain Burke, who were separated from us at Plymouth. Colonel Beach and Surgeon Mayer had been exchanged. Prison life had now commenced in earnest with us and we felt it deeply. We had heretofore been where we could see what was going on around us, and had the liberty of trading with hucksters and others. We were now cut off from the world, in a pen, with little or no shelter, and under the command of the most brutal, cruel, heartless, and inhuman men that this world produces. They were General Winder, Major Dick Turner, and Captain Tabb. As it is not the purpose of this history to record all the brutal acts of these men I give two or three to illustrate what fiends we were under. When General Stoneman made his attempt to rescue the prisoners, Winder issued an order called No. 13, which stamps the brute with infamy beyond redemption. In this order, which has been preserved, Winder commanded "the officers in charge of the artillery to open their batteries, loaded with grape-shot, as soon as the Federals approached within seven miles, and to continue the slaughter until every prisoner was exterminated." We had at this time six cannon bearing on us. "Was the prison mined," said Colonel Farnsworth to Turner, the jailor of Libby Prison, "when General Kilpatrick approached Richmond to attempt the rescue of the prisoners?" "Yes, and I would have blown you all to Hades before I would have suffered you to be rescued." Major Turner himself gave the prisoners to understand that if any more attempts were made for their rescue, the prison would be blown to atoms. The following atrocious order from rebel headquarters was afterwards issued. "Any soldier killing a Federal soldier, approaching the dead line, shall receive a furlough of sixty days; while for wounding one, he shall receive a furlough for thirty days." Under this order many were shot, who had no intentions of escaping. On May 29th, Assistant-Surgeon Nickerson was brought into prison. On June 10th, fifty of the officers, (Lieutenant-Colonel Burnham being one of the number,) were taken to Charleston and placed under the fire of the Union batteries on Morris Island.

We had religious services very regularly and they were well attended after the rebels found out that we would have them, cost what they would. I think it was the second service that we had, and on Sunday evening, for during the day word came to the commandant of the prison, Captain Tabb, that "the prisoners were praying for Abraham Lincoln, Grant, and Sherman." Services had commenced in the evening and the congregation were singing the first hymn, when in came Captain Tabb accompanied by the officer of the day. They at once inquired who had charge of the meeting, and being informed that it was Chaplain White, came up to him, and, entering into conversation, told him "he (Captain Tabb) could not allow any praying in there for the President of the United States." Meanwhile the hymn was concluded, when Chaplain Dixon, of the 16th, knowing what was up, at once stepped forward and began to pray, asking God to bless all in authority, especially the President of the United States, his cabinet and Congress, and all his advisers, also that he would bless General Grant and his glorious army; that he might be successful in capturing Richmond, the capital and stronghold of the rebellion; that he would also bless Sherman, spare his life, give him wisdom to carry out his plans, that his army might be a victorious one all through its campaign; that treason might be crushed, and traitors punished; that the time might speedily come when our dear old flag should wave over every village, town, and city of the United States, and we enjoy peace again. When the prayer was finished Captain Tabb withdrew from the crowd saying, "d——n smart prayer, but I don't believe it will amount to anything."

Many attempts to escape were planned, but few were successful. Major Pasco was caught tunneling, and the officer in command ordered him to fill up the tunnel. A revolver was held menacingly at his breast; but he resolutely refused. He was taken from prison and threatened with hanging, but was finally recommitted. On the 4th of July, the rebels gave us four roll-calls in the morning. An officer hoisted a small United States flag, which we all cheered lustily, much to the chagrin of the rebel guards. Then another sang the Star-Spangled Banner and we cheered that. Afterwards, without any preconcerted arrangement, we went into a large building, and held a meeting. Chaplain Dixon made one of the most patriotic prayers we ever heard. Then earnest, off-hand speeches were made, filled with a determination to persevere until the Rebellion should be crushed. Lieutenant-Colonel Thorp, 1st N.Y. Dragoons, was particularly defiant, and the audience showed their appreciation of it by frequent and prolonged applause. Of course such proceedings could not be tolerated by our rebel commandant, and he sent in his officer of the day to break up the meeting. The crowd quietly dispersed, after giving three rousing cheers each for President Lincoln and the little flag, the Proclamation, Grant, and Sherman.

To us it was a very satisfactory celebration. In the course of the afternoon the following order was posted on the bulletin-board:

SPECIAL ORDERS NO. 6.

C.S. Military Prison,
Macon, Ga., July 4th, 1864,

"I. Lieutenant-Colonel Thorp is relieved from duty as senior officer of prisoners for a violation of prison rules, and Lieutenant-Colonel McCrary will again assume that position.

"II. The same order and quiet will be observed on this day as on any other.

"III. A disregard of this order may subject offenders to unpleasant consequences.

"Geo. C. Gibbs, Captain Commanding."

After the meeting was over, Colonel Thorp was called out to headquarters, when the following conversation took place between Captain Gibbs and himself:

G.—"What's your name?"

Col.—"T.J. Thorp."

G.—"Were you addressing the officers in the prison?"

Col.—"I was."

G.—"What did you mean by it?"

Col.—"It was the desire of the officers that I should address them, which I did, as is the custom in our country on the 4th of July."

G.—"Sir, I shall put you in irons, and send you to jail."

Col.—"Very well, you can do so; but such treatment will not ameliorate my feelings toward you or the Confederacy in the least. We deem it not only a privilege, but a duty, to commemorate the 4th of July as the birth-day of a great nation, for whose defense and perpetuity we are willing to suffer, and die, if need be."

At this the Captain commuted his verdict to solitary confinement in jail without irons; but, before the guard arrived, the order was entirely revoked, and Colonel Thorp was sent back inside the stockade, with threats of summary treatment if he persisted in addressing the officers again on any subject.

SAVANNAH—CHARLESTON.

On July 28th, the first division of prisoners went to Charleston. This took of the 16th, Major Pasco, Quartermaster Robins, Captains Morse, Robinson, Burke, Hintz, and Lieutenant Bruns. The next day 600 more left for Savannah. In this squad all the remaining officers of the 16th went, they being Chaplain Dixon, Adjutant Clapp, Captain Turner, Lieutenants A.G. Case, Bowers, Strong, Andrus, Miller, Waters, Landon, and Blakeslee. On our way we busied ourselves by pitching the guards out of the cars when under full headway. Arriving at Savannah we were received by a large delegation of citizens, who were greatly interested, and wondered where our horns and tails were. Great was their surprise that we did not look different from their soldiers. The crowd was very great, and the police, aided by the city militia, could hardly clear the way for us to march through the streets. The officer in charge also was greatly confused, and gave so many wrong orders that it was a long time before we were able to march to the old United States Marine Hospital. We were confined in the yard surrounded on three sides by a brick wall eight feet high. While at this prison Lieutenant John M. Waters was taken sick with bilious fever. After a sickness of two weeks he was taken into the hospital on August 17th, dying the next day at 11 A.M. On the 19th, Chaplain Dixon was allowed to go out and perform the last duty of respect to our comrade in the presence of the Commander of the prison, Officer of the Day, Officer of the Guard, two Lieutenants, and four privates. Lieutenant Waters was very genial and, until his sickness, had kept up good courage. On September 2d, the Chaplains and Surgeons were sent to Charleston to be exchanged. This took Chaplain Dixon and Assistant Surgeon Nickerson of the 16th.

At an early hour on the morning of September 13th, we left Savannah and went to Charleston, where we were enthusiastically received and thrown into the yard of the jail. We here found Edward Woodford of Company I, who gave us some of the casualties of the enlisted men at Andersonville. He reported that the regiment stood it better than the other regiments who were captured at Plymouth, but already sixty had died. Two days after our arrival, Major Pasco, who was on parole at Roper hospital, (together with the balance of the 16th officers who left Macon in the first division, July 28th,) visited us, and through his efforts three days after, we joined him at Roper Hospital, by signing the following parole.

Charleston, S.C., C.S. America,

September, 1864.

"We, the undersigned, prisoners of war, confined in the city of Charleston, in the Confederate States of America, do pledge our parole individually as military men and men of honor, that we will not attempt to pass the lines which shall be established and guarded around our prison house; nor will we, by letter, word, or sign, hold any intercourse with parties beyond those lines, nor with those who may visit us, without authority. It is understood by us, that this parole is voluntary on our part, and given in consideration of privileges secured to us, by lessening the stringency of the guard, of free ingress and egress of the house and appointed grounds during the day, by which we secure a liberty of fresh air and exercise, grateful to comfort and health.

"Hereby we admit that this, our parole, binds us in letter and spirit, with no room for doubts or technicality of construction, and its violation will be an act of lasting disgrace. Signed:"

The firing upon the city was continued daily notwithstanding our being brought here and placed under fire of our own batteries as a means of saving the besieged city. Cheer after cheer went up as we heard the missiles crash through adjacent buildings.

On September 17th, at 1 P.M., a fire broke out near us. General Foster immediately opened on the place, the fourth shell bursting in a dwelling house near by, setting that also on fire. So accurate were Gilmore's guns aimed that two shells burst in the burning buildings; and the negroes told us that one struck an engine, entirely destroying it, and killing several firemen. One shell struck our wall, tearing it nearly down. Several pieces of shell came into the yard where we were, but no one was injured. One of our officers in Roper Hospital was wounded in the arm by a piece of shell that came through the roof while he was eating dinner, making a flesh wound, smashing up the table, and passing on through the floor. In the evening we often watched the shells coming, and it was a beautiful sight. We first caught the flash of our guns; then, after waiting a few seconds, we could see the missile, which looked like a shooting star, climbing up higher and higher; and, when it reached its full hight, we could hear the report of the guns that sent it; by this time the shell was so near, that we could hear its sharp, shrill shriek; then it gradually descended, and approached until it was right over our heads, when it burst, most of the pieces going beyond us. The report of the explosion then reached us, as we heard the fragments rattle among the brick walls, or crash fearfully through the wooden tenements. Shells from Gilmore's batteries on Morris Island were continually making sad havoc in the city, killing and wounding from one to eight persons daily, and going through entire blocks. Sometimes a single percussion shell would blow up an entire building.

On October 2d, Major Pasco received a letter from Lieutenant Colonel Burnham, stating that he was once more a prisoner of war and confined at Wilmington. This was news that very much surprised us. It seems that after being exchanged at Charleston, he had been home and had got newly fitted out, and was on his way to the regiment. At this time Co. "H" and a few others of the regiment which had been gathered from different points were stationed at Roanoke Island, and this composed the 16th Connecticut Volunteers, on the Union side of the lines. Lieutenant Colonel Burnham took the steamer "Fawn" at Norfolk, and while passing Coinjock, on the Albemarle and Chesapeake canal, they were fired into by guerrillas, who lay concealed near the shore. One was killed outright and several wounded. The remainder were taken prisoners, and the steamer was burned. The prisoners were marched to Elizabeth City, and turned over to the rebel authorities. Colonel Burnham was soon after again paroled.

COLUMBIA.

On October 5th, we went to Columbia. Captains Morse, Burke, and Robinson escaped by jumping from the train, but were recaptured on the 11th, and rejoined us next day at Camp Sorghum.

That we might get wood the commandant allowed a certain number to be paroled each day. Their names were written on a piece of paper, handed to the officer of the day, who instructed the guard to pass them in and out at their pleasure till night. On the 3d of November, Captains Robinson, Dickerson, and Burke, took advantage of this, and successfully escaped, an account of which has been written by Captain Dickerson as follows:

"Upon the evening of the 3d of November, 1864, a large number of officers were paroled for the purpose of bringing wood for cooking and building purposes. Three of us,—Captains Burke, Robinson, and myself—finished getting wood, saw our paroles destroyed, returned to camp, and, after a short time, recrossed the guard-line, and, unsuspected, made our way to the woods. We kept slowly through the woods, until we were safely beyond the reach of the rebels. Night coming on, we traveled through the fields and woods, until we struck a road which ran parallel with the Congaree River; and traveled southward about twelve miles. We then left the public road, and traveled through the woods, toward the river, which was about eight miles distant. Upon the afternoon of the 4th, we fell in with five other of our escaped prisoners from the same camp, who escaped two nights previous to ourselves by bribing the guard. They had been in the swamp two days, and had made an arrangement with the negroes for provisions, and information where two boats could be found. We joined our parties, and in the evening the negroes met us, and furnished us with the promised necessaries. Proceeding to the river, which was about two miles distant, we found the boats which accommodated all of us. At about eleven o'clock, all things being ready we embarked in our boats, and paddled down the river. Early in the morning, we landed, having secured a hiding-place for our boats and ourselves, and devoted the day to cooking and sleep. Darkness again found us upon the river; and we soon passed the bridge of the South Carolina railroad, which was guarded by sentries; paddled all night; and again secured ourselves for the day.

"While engaged, some in cooking, others in sleep, our attention was attracted by loud talking; and soon we discovered a boat below us, upon the river, being poled up the river by negroes. One of our party posted himself upon the bank of the river, where he could hail the boat without discovering the presence of the rest of the party, and, hailing the negro in charge of the boat, informed him that he was in want of provisions. The negro replied that he had no time to stop, and persisted in keeping on his course. When all entreaty failed, the captain told him he was a 'Yankee' officer, escaped from rebel prison, making his way north and needed provisions, and information regarding obstructions in the river. The negro no sooner learned his true character than he immediately landed, secured his boat, staid with us all day, cooking rations, and giving very valuable information. He claimed in return for his service, a sight of the rest of the party, whom he had instinctive shrewdness to know were concealed close by. Night again found us upon the river; and after a few hours' sail we landed at a farm, of which we had been informed by the negroes, where we succeeded in capturing a goat, which supplied us with meat to our journey's end.

"Monday and Tuesday nights nothing occurred to delay our progress toward the desired goal. Wednesday evening, we had been going down the river but a short time, when we reached a ferry, where we made the acquaintance of four negroes, who furnished us with a large quantity of sweet-potatoes, salt, and meat. Upon parting with them, they bade us God-speed and a safe journey. Elated and happy with our success, we kept steadily forward; and soon after we landed to again consult the negroes, who told us we were but five miles from a battery, mounting two pieces, upon the right bank of the river, guarded by rebel soldiers. After receiving other information and provisions, we parted company with the last of our negro friends, and proceeded down the river, passed the battery in safety, and, landing, waited for the darkness of the night to finish our journey to the coast. Thursday night, full of hope, we again took to our boats, and meeting no interference, reached the coast safely.

"The light of Friday morning, November 11th, revealed to us the spars of a ship, which we soon made out to be one of the United States blockade. It was the Canandaigua, Captain Harrison; and after sundry attempts we succeeded in getting safe on board, under the protection of the star-spangled banner. We came north on the Fulton."

Of the many devices for escape, all were not as successful as the above, as our major will testify.

Major Pasco had become possessed of two gold dollars which had been sent him in cans of solidified milk from the north. He had bribed a guard to permit Captains Morse, Turner, and himself, to pass out that evening between the hours of eight and ten, for the two gold dollars and a jack knife. At the appointed time, and all three officers being ready, they approached the dead line, and Captain Morse going first gave the booty to the guard who allowed him to pass and also Captain Turner. At this point an officer who was lying on the ground and trying to escape on his own hook in another manner, jumped up and passed out. The guard had passed out three according to agreement. Major Pasco came along and was halted by the guard, and the following conversation took place:

Major. You agreed to pass out three of us.

Guard. Well I have let out three. You must go back.

Major. That last man you let pass didn't belong to my party.

Guard. Can't help it, I have let out three and you must go back.

The Major undertook to explain matters, but the sentinels along the line commenced to fire at him and he beat a hasty retreat into camp, where he dreamed all night of the officer who euchred him out of an escape.

But the Major was not to be foiled out of a successful escape. He made a second attempt to reach the Union lines by writing out a false sick certificate to pass to the hospital, but he was recaptured after seven days absence. The third time he succeeded. Shortly after the train left Columbia, while we were going to Charlotte, N.C., he jumped off the train and was overtaken by Sherman's army. Lieutenant Bruns also escaped from this prison, but was recaptured and brought back after an absence of ten days. Captains Morse and Turner were returned to prison, having been absent a month. There were so many escapes from this prison that on the 12th of December, the rebels marched us into the city and confined us in the yard of the Insane Asylum with a brick wall around it eight or ten feet high, and eighteen inches thick.

CHARLOTTE, RALEIGH, GOLDSBORO, WILMINGTON.

The month of February was full of rumors in regard to our exchange, but it was an old story to us given out to prevent our trying to escape. On the 14th and 15th we were moved to Charlotte. General Sherman had by this time arrived within two miles of the city and was posting his artillery on the hill, (Camp Sorghum[3]) where we had been imprisoned a few weeks before.

As the train started. General Sherman opened his batteries on the city. Lieutenant Landon and a party of sixteen had managed to hide themselves between the rafters under the floor of the second story of the Hospital building, where they stayed with little or no food for forty-eight hours and fell into the hands of Sherman. Major Pasco, Captains Morse and Turner, who jumped from the first train that left Columbia, secreted themselves in the woods and General Sherman overtook them on his way north. The remaining officers of the 16th were sent to Charlotte with the rest of the prisoners, arriving there on the evening of the 16th, having been delayed some hours at a point 40 miles from Columbia by running into a drove of cattle which the rebels were driving ahead of Sherman's course. The engine, tender, and one car was thrown from the track, killing three beef creatures. This was issued to us, being the first meat of any kind that we had tasted for four months and ten days.

Disembarking from the cars on the morning of the 17th, we marched three-quarters of a mile to a little pine grove, which we called Camp Necessity or Camp Bacon, by reason of some bacon

[3] So named, because that was the principal ration we received while there.

having been issued,—the only meat rations for over four months. The fact that we were soon to be exchanged was now confirmed by the rebels taking off part of the guard and allowing them to use unloaded muskets. Still[4] "old fish" turned away and would not believe it, and a good many escaped, as it was a very easy matter. Captain Hintz and Lieutenant Bruns immediately left, and Lieutenant Johnson with Lieutenant Eli P. Alexander, (Co. H. 26th Mich.,) went out about noon, and nothing has ever been heard from them. It is supposed that they were murdered by guerrillas. On the 20th, we signed a parole not to serve against the Confederate States until regularly exchanged, and that evening we left for Raleigh, where we stayed a few days and thence proceeded to Goldsboro, where we were quartered in the Court House. On the 27th we signed the following parole.

We, the undersigned, prisoners of war, do give our Parole of Honor, that we will not take up arms again, nor serve as Military Police or Constabulary force in any fort, garrison, or fieldwork, nor as guards of prisoners, depots or stores, nor to discharge any duty usually performed by soldiers: until exchanged under the provisions of the cartel entered into July 22d, 1862.

About six o'clock in the evening of the 28th, we left Goldsboro by cars crowded both in and outside, amid cheers and songs, bound for Wilmington, which was now in our possession. Riding all night, daylight found us standing on the track at Rocky Point, a few miles from Northeast Bridge, and fourteen miles from Wilmington. This was the picket line and out-post of the rebels. At nine o'clock Colonel Hatch on a special train arrived and proceeded ahead with

[4] The first six months of prison life one is called a "fresh fish," the next four months a "sucker," the next two a "dry cod," and the balance of his time a "dried herring," or "old fish." After exchange he becomes a "pickled sardine."

As soon as a new prisoner made his appearance at the gate, the cry of "fresh-fish," by each one ran through the prison, and a general rush was made for the gate. An eager group surrounded him, and while some would be seriously asking questions concerning his capture and listening to his pitiful story others would call out "Take your hands out of his haversack;" "Give him air;" "Keep that louse off him;" "Don't take his clothes;" etc. All this affected them strangely at first, and produced a hearty laugh for us.

a flag of truce on the engine. We soon started and reached neutral ground about ten o'clock. About a mile past Marlboro, we came in sight of the "Boys in Blue." Three cheers were immediately given. As the train came to a stand still, all were silent, and we quietly disembarked and were counted through the ranks of twenty Union soldiers, they presenting arms. As soon as we were able clearly to comprehend that there was not somebody at our side with a loaded musket, cheer upon cheer was given, caps were thrown high in the air, some were hugging and kissing each other, and others cried. Whilst some sang, others were laughing and rolling on the ground. Finally our old pans, kettles, bags of meal, and bundles of all sorts were flung high in the air. Cheer upon cheer was given for Lincoln, Grant, Sherman, and General Exchange. Some hugged the horse of a Colonel, who seemed to have command over us, but tried in vain to get us into line. Taking up the line of march, a mile brought us to Cape Fear River, and in sight of the United States flag. Cheer after cheer ran the whole length of the line. The 6th Connecticut was encamped on the bank of the river, and at the end of the pontoon bridge which we had to cross, they had erected a handsome arbor decked with flags. In the center of the arch, surrounded with a wreath of evergreens, were the words

"WELCOME, BROTHERS."

A band was placed here, and played "Hail to the Chief who in Triumph Advances." Some few cheered but more were moved to tears, and kissed that dear old flag which they had so long suffered for, and all marched under with uncovered heads. Arriving at the top of the hill, we found the whole division of troops in line to receive us. On reaching camp we partook of a breakfast of United States rations, the first full meal which we had had for a long time. At noon those that were able walked to Wilmington, a distance of nine miles. We were here met by officers who provided us with another meal, and then we were quartered in churches and private residences. The next day (March 2d,) we took the steamer "General Sedgwick" for Annapolis, Md., where we remained a week, receiving two months pay and a leave of absence for 30 days, at the end of which time we were to report at Camp Parole, Annapolis, Md.

Extracts from a statement made in 1867 by Lieut. B.F. Blakeslee, to a "Congressional Committee on the Treatment of Prisoners of War and Union Citizens;" John P.C. Shanks of Indiana, Chairman.

At Macon, for quarters we had merely a roof which sheltered us from rain; but no protection from the cold damp nights. Rations consisted per diem of a pint to a pint and a half of corn-meal; from poor to fair in quality, and occasionally in lieu thereof baked corn-bread from one to two inches thick, three inches long and three inches in width, containing a large quantity of water and very hard; rice half a tea-cupful, or in lieu thereof the same quantity of beans; salt, a teaspoonful; bacon, from three to four ounces; invariably decayed and maggoty, with only two exceptions, two issues of soap were made and that was the end of trying to keep clean; meat of any kind except the bacon above mentioned we never saw. No cooking utensils were issued. The rations of wood were not sufficient to do our cooking. From the 11th to the 18th of July, none whatever was issued, and rations had to be eaten raw.

This caused much hunger, suffering, and misery, as the beans and rice could not be eaten. The quality of the water in this prison was good and usually plenty, though some days the supply was short. Our treatment here was generally bad. The length of confinement was three months.

At Savannah we were better treated in every respect than in any other prison, provided with tents, and cooking utensils, and a good supply of rations of good quality. Fresh beef was issued nearly every day. The water though, was very poor; having a fetid smell, and unpleasant taste, and could only be used for drinking purposes by filtering through charcoal; or burnt rice or meal steeped in it. The length of confinement was six weeks.

At Charleston Jail Yard, only a part were provided with shelter; and in rainy weather the yard was flooded with water, so we could neither lie down nor do any cooking. In pleasant weather, it was as hot as an oven. Little or no fresh air could come within those walls which were twelve feet high, and in addition surrounded by buildings. When there was a breeze, there were whirlwinds of dust which would almost suffocate us. The water was very poor, making a great many sick. The rations consisted of small quantities of corn-meal and rice, and one ounce of bacon per day; but after a week or so the bacon was dispensed with. There were no utensils for cooking, and but little salt was issued. In addition to our deplorable condition we were under fire from Gilmore's batteries, whose shells were continually bursting around us, occasionally coming amongst us and twice tearing the wall away. On one occasion for sixty hours we had not a morsel to eat.

At Roper Hospital Prison, in Charleston, our exact rations were for ten days, two and a third quarts of corn-meal, two quarts of rice, three pints of black beans (including bugs,) and four ounces (daily) of fresh beef, or in lieu thereof, two ounces of bacon. No cooking utensils were to be had. We certainly should have starved to death here, had it not been for an arrangement made for obtaining money which enabled us to purchase food of the citizens. The authorities gave us Confederate money in exchange for our drafts (in gold) on the North.

At Columbia we were turned into an open field like a drove of cattle to pass the winter months, without any shelter whatever, neither cooking utensils, axes, spades or anything were issued that would enable us to make ourselves comfortable. With scanty clothing, but few blankets, some without shoes, we were left here to pass the winter as best we could. Rations consisted of corn-meal and rice. Twelve days rations of rice made one meal. Salt was issued in small quantities, and for four consecutive days we had none at all. On September 26th and 27th, we had nothing to eat. One or two issues of flour were made, but no meat of any kind was provided.

While on our way to Charlotte, the train ran into a drove of cattle, killing three, which were issued to us, making the only meat rations we had had for four months and ten days. At Raleigh we met several trains loaded with enlisted men going north to be paroled. Nearly all were sick and very dirty and black; no soap having been issued to them for six months. Nineteen out of one train had died since leaving Salisbury, a distance of 132 miles, mostly of starvation, though some who were on the top of the cars were frozen to death. On Monday morning they had half a loaf of bread, weighing not over five ounces issued to them, since which time the rebels had issued not a particle of food. It was then Thursday noon. How soon thereafter rations were issued to them I am unable to state.

As soon as a new prisoner made his appearance at the gate, the cry of "fresh-fish," by each one ran through the prison, and a general rush was made for the gate. An eager group surrounded him, and while some would be seriously asking questions concerning his capture and listening to his pitiful story others would call out "Take your hands out of his haversack;" "Give him air;" "Keep that louse off him;" "Don't take his clothes;" etc. All this affected them strangely at first, and produced a hearty laugh for us.

CHAPTER VIII.

PRISON LIFE OF THE ENLISTED MEN.

It is to the credit of the rebel soldiers whose good fortune it was to capture our command, that we were treated with considerable courtesy and kindness while in their power. Our men were allowed to retain their blankets and overcoats, and all little articles of value which they might have upon their persons. Many of the men had about them large sums of money which they were allowed to keep. From Plymouth, the long and wearisome march made to Tarboro (an account of which is given in the preceding chapter,) together with scanty rations and exposure, told severely on the men, and many were sick and feeble; and it was with no little pleasure that, on the morning of April 29th, they marched to the depot in the town to take cars to Camp Sumter, where, as the rebels informed them, rations would be dealt out plenteously. They were crowded aboard small box cars by forties, and, in addition, six rebel guards were stationed in each car, occupying the door. Of course under such circumstances, they were nearly suffocated, and were pressed almost out of shape. The train started at 10 o'clock, stopping at Goldsboro, where rations were issued, consisting of three small hard crackers and a little scrap of bacon, to subsist on for the next twenty-four hours. Although arriving at midnight at Wilmington, they were not allowed to get out of the wretched cars until morning. At sunrise they were marched down to the dock, and conveyed by ferry boats to the opposite side. Taking the train in waiting for them, they proceeded to Charleston, arriving there on Sunday morning, May 1st. In the afternoon they were transferred to another train and put aboard platform cars and at a rapid rate went to Savannah, Georgia. But before reaching there they were overtaken by a storm and thoroughly drenched with rain. Changing cars at Savannah, they proceeded to Macon, and thence to Andersonville, arriving there at nine in the evening. Leaving the cars they were marched into an open field near by, where they remained during the night, and marched into the prison pen the next morning under the escort of a strong guard. How each one felt as he entered this "hell upon earth," can little be imagined. The first night ten died near the position of the 16th. The men seemed to stand it pretty well at first, much better than the other regiments captured at Plymouth, and it was not until the 20th of June that the first of their number died, Alonzo A. Bosworth, Co. D. But by the 1st of August, some of the Sixteenth died nearly every day.

The inhuman treatment which our men experienced in Southern Prisons has been told over and over, and is well known in history and need not be repeated; but this history would not be complete without inserting the following testimony of rebel barbarity taken from the diary of Corporal Charles G. Lee, (Co. B.,) who died from exposure and lack of food, immediately after being exchanged at Wilmington, N.C. He writes as follows, "Again I am called to bid adieu to the passing year, but under very different circumstances from any in which I have ever been. During the year 1864, I have passed eight months in the most degrading imprisonment. In that time, our inhuman captors had not furnished shelter of any kind; and we have repeatedly been for two and three days at a time without a morsel of food; and even that we have received would at home have been generally thought unfit for swine. We have not had a particle of meat for forty-two days, and but little molasses, or any thing to take the place of it. Our rations chiefly consist of about a pint and a half of coarse corn-meal, and half a teaspoonful of salt daily. Now and then we receive a few beans or sweet potatoes. Many a night have I lain awake because I was so hungry that I could not sleep."

About the 1st of September the prisoners were removed to Charleston, South Carolina, where they remained about four weeks, when the yellow fever broke out and raged so fearfully among the rebel forces who guarded the prisoners, that they were removed to Florence, where they spent the winter months. During the latter part of December, 1864, and the months of January and February, 1865, the men were—a few at a time—paroled and allowed to come north, and afterwards were regularly exchanged, thus ending the career of the Sixteenth in prison, with the heavy loss of over fifty per cent. in deaths, in a period of a few months. A more detailed account has been published by Sergeant Major Robert H. Kellogg, in his "Life and Death in Rebel Prisons." Among the number who escaped from prison, were Quartermaster Sergeant Hiram Buckingham and Andrew J. Spring, of Company K. An order was received for the names of all sailors at Andersonville. Sergeant Buckingham suspecting it was for the purpose of exchange, obtained a suit of sailor's clothes, and accordingly took the name of Johnny Sullivan, a sailor who had died in the hospital a few months before. In about a week after the names had been registered, the sailors were ordered out of prison. Buckingham answering "Here," to the name of Johnny Sullivan, passed out without detection. They went to Charleston, thence to Richmond, and were exchanged, having been in prison just six months.

Andrew J. Spring in some manner procured money enough to bribe a guard, who allowed him to escape with two comrades. They were five days in reaching the Union lines, living meanwhile on sugar-cane, green corn, and persimmons. Traveling in the woods, they guided themselves by the moss, which grows heaviest on the north side of the trees, and successfully passed three lines of rebel pickets.

The shooting of prisoners who came near the "dead line," was of almost daily occurrence; for if they were near it with no intention of escaping, the sentinels would fire. The regiment lost one man in this manner, William Drake of Company A, who was shot December 4th, 1864.

CHAPTER IX.
1865.
NEW BERNE,—HARTFORD, CONN.

Company "H," (Captain Barnum,) who escaped capture at Plymouth, by being detached and sent to Roanoke Island for duty in April, 1864, was reinforced now and then by men who had previously been detached for special service, or were absent sick, also by a few who were exchanged from time to time, representing every company, and this composed the 16th regiment in actual service. Captain Barnum labored with much zeal under many difficulties, to preserve the former prestige of the regiment. During December the regiment proceeded to Plymouth, and went thence on an expedition to Poster's Mills, about ten miles, destroying the mills and a large quantity of grain, and returning with various spoils. On another occasion the regiment went to Hertford, where they captured large quantities of cotton, tobacco, finished carriages, and buggies, several thousand feet of lumber, several mules, and forty contrabands. And again one bright night Captain Pomeroy with sixty men proceeded by steamer up the Alligator river, capturing a barge and three small sail vessels containing twenty-five hundred bushels of shelled corn, together with the outfit of fifteen men with their mules and carts. They were intending to take the corn to a mill near by to be ground. The regiment also made several unimportant raids to Columbia, Edenton, and the adjoining country, until March 4th, 1865, when they were ordered to New Berne, N.C., where the exchanged prisoners joined them and remained on provost duty. Most of the officers were quartered in the houses at the corner of Craven and Union streets. Colonel Beach having been released from Libby Prison in May, 1864, was assigned to various duties in Washington, only once rejoining what remained of the regiment. That was at New Berne, where he was taken sick and soon departed on sick-leave.

Colonel Frank Beach was a graduate of West Point Academy, class '57. He was stationed at first at Fortress Monroe, as a brevet second lieutenant of artillery.

At a later date he was ordered to the far west with General Gibbon, and took part in the well-known Utah expedition in 1858. The sufferings of that campaign and the winter encampment on the prairie were shared by him, as well as the almost unendurable ennui of later days, when Digger Indians or inimical Mormons were the only society accessible to the small garrison.

When the war broke out Colonel Beach was post adjutant at Port McHenry near Baltimore, and remained in that position for some time. He took some share in McClellan's advance, and was stationed at Yorktown as an officer of artillery. But in the summer of 1862, he was permitted, by special order of the war department, to accept the colonelcy of the Sixteenth Connecticut regiment which had been tendered him by Governor Buckingham. He commanded the regiment at the battle of Antietam, showing great personal bravery and heroism during the engagement. He galloped hither and thither on his white horse over the field, trying in vain to draw the men out of the desperate charge into which they had been ordered, and sad and full of woe was his heart on the night after the struggle, when the broken remnants of the Sixteenth gathered around him in the rear of the battle ground. He made personal inquiry after each of the wounded, and visited a number of them on that evening and the following days, doing for them all that was possible.

The winter which followed made him an invalid with a disease whose seeds had been laid in the Utah campaign. But, as he was reluctant to leave the regiment, he accompanied it in an ambulance on the long marches down Virginia to Fredericksburg. With him, and sharing the same ambulance, was Colonel Griffin Stedman, the heroic commander of the Eleventh Connecticut, still lame from Antietam wounds. They became firm friends, and not unfrequently in those cold evenings the ambulance would harbor a merry party, which, by the light of a hospital lantern, and in the sight of the surrounding camp fires, would speed the long hours by merry conversation. Major Converse, Adjutant Barnum, (both fallen) and Dr. Mayer would bear them company.

The greater part of that winter the Colonel remained with the regiment, but was finally forced to take sick leave. He returned to it in the summer at Portsmouth, Va., and held command during the siege of Suffolk, and the charge on Longstreet's army. Then he conducted it to North Carolina, where he remained in command of a brigade, until at Plymouth, he was taken prisoner with the regiment and all the other troops that garrisoned this surprised outpost.

After the war Colonel Beach was for some time in command of a solitary fort near Washington. He was soon after stationed at Washington, and then at Fort McHenry. His old trouble having reappeared with more than its former violence and persistency, he was placed on the retired list, and endeavored to regain his health, but with only temporary success. He died at New York, in the New York hotel, on Wednesday evening, February 5th, 1873.

Colonel Beach was a gentleman of very handsome appearance and strong masculinity of deportment. He was widely and well read, and as thoroughly acquainted with the progress of modern philosophy and science as with the prominent poets and writers of belles lettres of all ages. He had an elegant yet terse method of expression, and a flashing quality of wit. But no man was of kinder heart, and in the regular army his good nature had become proverbial. In his first connection with the Sixteenth Connecticut Regiment under unfortunate circumstances, many misunderstandings between him and the men gained ground. This, as in some other regiments, was owing to the jealousy with which the volunteer soldiers, fresh from home, regarded regular army officers, and to the disagreeable impression the necessities of army discipline made on them. But, a little later, and at the close of the war, there was not a man of the regiment who was not warmly attached to the Colonel, admired him, was proud of his bravery, his military knowledge, bearing, and of his standing in the army. "Little Moustache," and "Black Eye," the men used to call him among themselves, and they made a boast of him to those of other regiments. He was as splendidly endowed with all the qualities that make the true and noble man, as with all those that please and captivate in society. For years a sufferer from a harassing disease, yet few came in personal contact with him but will regret his demise as that of a person of fine and polished intellect, and engaging manners, and of a great-hearted gentleman.

Surgeon Mayer who was exchanged in May, 1864, was ordered to the Foster General Hospital at New Berne, N.C. There he remained in charge of four wards until the latter part of September, when the Chief Surgeon of the hospital went north, and he succeeded to the charge of the institution. Immediately afterwards the yellow fever broke out. Its ravages in the city of New Berne and among the garrison are a matter of general history. There were only a few of the Sixteenth at New Berne at the time, and most of these had been detailed as clerks or nurses to the hospital, at Surgeon Mayer's suggestion. Jasper A. Winslow, Company "C," who at his own request, through the Surgeon's influence, was ordered there as clerk, took sick at once, and died in a few days. W. Chester Case, Company "H," was doing clerk's duty and proved very efficient at this terrible time. He held out courageously, and kept the reports of the dying, of their places of burial, of their possessions and accounts, until he himself was seized with the fever. When it is considered that sometimes as many as thirty or forty died in one day at the hospital, an idea may be formed in regard to the difficulty and labor of keeping reports. Under Surgeon Mayer's personal treatment Case and a few other Sixteenth men, sick at New Berne, recovered. But at last he took sick himself. For two days it was doubtful whether he would live. Then, some favorable symptoms occurred, and Medical Director Hand sent him to Morehead City. After a two weeks convalescence, he returned to New Berne, where Surgeon Rice and Surgeon Cowgill, who had been in charge since his sickness, lay also attacked by the fever. He took charge again and so continued until his appointment to a different office. During the epidemic, eighteen assistant Surgeons had shared his labors, all of whom had in succession been attacked by the fever, which carried off nine of them.

General Palmer, in recognition of Surgeon Mayer's services, appointed him Medical Purveyor of the district, and this office he held until the muster out of the regiment. A complimentary order was issued to him. His management of the hospital during the yellow fever time, and his administration of the Medical Purveyor's department, met with general praise.

The Doctor says: "I got out of the hospital where over five hundred died, and had saved seventy patients above the general average; and I got out of the Medical Purveyor's office, where I had some million dollars worth of property to administer and was square with the United States Government, all but fourteen pounds of nails, which I couldn't account for; so I may consider myself fortunate. But the Quartermaster's Department kept writing for those nails every quarter for four or five years."

In May it was generally conceded that the war was ended, and Captains Dickerson and Turner resigned and went home. Adjutant Clapp also resigned while home on leave of absence, and Lieutenant Landon was made Adjutant. June 19th, Major Pasco returned from Connecticut, with the necessary muster rolls and papers to enable us to complete the muster-out rolls. Then both day and night did the officers work on discharge papers and muster-out rolls. Finally on Saturday, June 24th, the rolls were examined and we were honorably mustered out of the service, at 5. P.M., by Captain John D. Parker, A.C.M., Second Massachusetts heavy artillery, the men remarking, while standing in line, waiting patiently; "that while it did not take long to enlist, it took a long time to get mustered out." It proved quite true; for while we enlisted for three years, it was not supposed that we should be out more than three or six months at the most; and many of the men enlisted expecting to return in a short time, not one of us realizing the hardships and sufferings we must pass through. But who of us regrets the faithful service performed for our country. How many around us to-day do we see who blush and say the greatest mistake they ever made was that they did not go to the war. How many would say as did a prominent man to me, the day we returned home; "I would give fifty thousand dollars to have seen and been through what you have."

The regiment participated in the following

ENGAGEMENTS.

Antietam, Md.—September 17th, 1862.

Loss in killed, four commissioned officers, 38 enlisted men; wounded, eight commissioned officers, 176 enlisted men; captured, 12 enlisted men; Total Loss, 238.

Fredericksburg, Virginia.—December 12, 13, and 14, 1862.

Loss in wounded, one enlisted man. Total loss one.

Edenton Road, Suffolk, Virginia.—April 24, 1863.

Loss in killed, one enlisted man; wounded, seven enlisted men. Total loss, eight.

Providence Church Road, Suffolk, Virginia.—May 3, 1863.

Loss in killed, two enlisted men; wounded, one commissioned officer, seven enlisted men. Total loss, 10.

Plymouth, North Carolina.—April 20, 1864.

Loss in killed, one enlisted man; wounded, one commissioned officer, 11 enlisted men; captured, 23 commissioned officers, 400 enlisted men. Total loss 436.

CASUALTIES.

Killed in action,	47
Died of wounds,	45
Died of disease,	73
Died in prison,	177
Shot in prison,	1
Supposed shot while attempting to escape from prison,	1
Lost at sea from burning steamship "General Lyon,"	1

Drowned from accident,	1
Drowned from collision of "Black Diamond," on Potomac river	7
Total deaths,	353
Discharged prior to muster-out of the regiment,	386
Captured at Antietam,	12
Captured at Plymouth,	435
Wounded,	212
Missing,	56
Total Casualties,	1454
The regiment as mustered into service numbered,	1007
Recruits,	75
Officers appointed since first muster,	5
Total,	1087
Number returning home with the regiment,	131

The officers who returned were:

Lieutenant Colonel John H. Burnham, Commanding.

Major Henry L. Pasco.

Surgeon Nathan Mayer.

Quartermaster Gordon Robins, Jr.

Adjutant Herbert Landon.

Chaplain Charles Dixon.

Captains C.W. Morse, Henry Hintz, Joseph H. Barnum.

Lieutenants George A. Bowers, A.J. Case, Harmy Bruns and B.F. Blakeslee.

The above list of casualties is far from correct, but is as accurate as can be obtained from the Catalogue of Connecticut Soldiers, issued by the Adjutant General of the state, with some additions well known by surviving members. The casualties were greater rather than less, than the figures given above. The number given as killed, are those who were killed outright; but it is generally believed that a greater number died from wounds than the regiment has been credited with, for within forty-eight hours after the battle of Antietam, nine died from wounds; and out of the seven wounded at the fight on Providence Church road at Suffolk, three died. It is therefore safe to say that forty-five is too low a number of deaths out of 212 wounded. It is also believed that more than twelve were captured at the battle of Antietam. There is also no doubt that the regiment lost at least two hundred and twenty, by death from rebel cruelties and starvation. Eighteen can be counted who died immediately after arriving at Wilmington and Annapolis from southern prisons.

The Sixteenth was always called an unfortunate regiment; for if there was any special hardship to endure, the regiment was sure to be called on to experience it, either by accident or otherwise. It was our bad luck.

The list of casualties show in how many ways the men were lost. Some of the men and two of the officers never could be accounted for. Lieutenant George Johnson, who escaped from prison at Charlotte, has never been heard from. It is supposed that he was shot by guerrillas, who infested the locality in which he said he was going to reach the Union lines. Lieut. William H. Miller was last seen at Washington on his way to the regiment after having been exchanged. It is supposed that he was taken sick and died in the hospital. Elbert Sutleff, Co. K, was lost at sea by the burning of the steamship "General Lyon." Quite a number of the men who were returning to the regiment after captivity, were on a steamer going down the Potomac river in the night when the boat collided with the "Black Diamond." All the passengers jumped aboard the Diamond, which went down, and the regiment lost seven men. Some of the saved remained in the water three or four hours.

On Sunday, June 25th, the regiment proceeded to Morehead City, and embarked on the steam transport "General Meigs," with the 23d Massachusetts Regiment, and the Connecticut Brigade Band, which, learning that we were mustered out and on our way home, had obtained an order from General Palmer to be sent home under charge of Lieutenant Colonel Burnham. The sea being rough we did not leave till the next morning. We arrived at New York late on the afternoon of the 28th. Taking on some rations we proceeded to New Haven, arriving there at an early hour in the morning. Procuring special transportation, we reached Hartford at eight o'clock. Being a Hartford regiment, the citizens had made extensive arrangements to welcome us, but the 18th Connecticut had arrived that morning by boat and while their attention was drawn towards them, the Sixteenth suddenly entered the Asylum street depot. The news soon reached the State House Square, and the bell was rung, and by the time the regiment was in line the Governor's Guard, City Guard, and Colt's Band were on the "double quick," and thousands of citizens were hurriedly approaching the depot. Many who had seen the regiment leave for the war three years before, and now witnessed the decimated ranks were effected to tears. Those who had husbands, brothers, or relatives in the regiment, watched us eagerly and looked strangely into the ranks, hardly believing that any could be missing. One lady, the wife of an officer, was told for the first time of her husband's death. So great was her grief, that friends who accompanied her could hardly get her into a carriage to convey her home.

From the Hartford *Courant*, June 30th, 1865.

THE SIXTEENTH.

"Everybody supposed that this gallant regiment would arrive here on the 9.45 regular morning accommodation train from New Haven, but the "boys" being accustomed to making surprises secured special transportation and were in the Asylum street depot by eight o'clock. Here they were received by the Governor's Guard, the City Guard, and Colt's band, and escorted up High to North

Main street, down Main to State. On the march, though few in numbers, (but one hundred and thirty enlisted men returning) their tidy and soldierly appearance was the subject of general comment. Being a Hartford regiment there was an unusual interest manifested to see them, and signs of welcome were apparent on every hand. Arriving in front of the United States Hotel, they were drawn up in line, and Governor Buckingham made a brief speech congratulating them on their safe return and extending them cordial greetings on behalf of the State."

Hon. Ezra Hall was then introduced, and welcomed the regiment in behalf of the city in an eloquent manner as follows:

Officers and Soldiers of the Sixteenth Connecticut: Heroes of many a hard-fought battle, and worthy veterans of a redeemed country! On a beautiful summer day in 1862, when the nation was in its greatest peril, and rebel powers seemed successful for a time in driving back the armies of the Republic, placing in jeopardy all our hopes and every interest of free government, you, more than a thousand strong and valiant men, volunteered to place your names upon the muster-roll among the country's defenders. That roll of honor will go down the centuries and in the far off future the lover of freedom will unroll the scroll, and call your names as among the heroic volunteer force who cemented the Union of these States, and proclaimed through the immortal Lincoln, freedom to all mankind. It was hard to leave your situations, your homes, and those you loved. And a sharper pang would steal along your feelings as you thought the step might take you forever from the dear New England hills and all you held dear. But manfully you resolved, and the pensive feelings that evidenced so well your better life, while preparing for the decision, made you braver as you stood on the enemy's grounds, striking for the very homes you left, and the government of your fathers. But the question was decided for country; and you went out from your homes to camp. Hard were your pillows, and the long and weary days went slowly on. Friends gathered to see you in your tents, and to pay you a tender tribute ere you went out to battle. They chatted and counselled with heavy hearts, but asked holy benedictions for your safety and through you for the country. Who of you can forget the stirring emotions of those meetings and partings. For a few days you drilled and disciplined for the coming strife. The order came. You struck your tents—passed down the very streets on which we stand, with colors flying and music measuring out the solemn step of war. No braver regiment ever went out from our city or State. Made up mostly of Hartford men, born and cradled under the very branches of the Charter Oak, we knew well what would be your history, and we watched with pride your firm and steady ranks, as you filed along these streets. As we followed with anxious eye the steamer winding down the Connecticut, bearing its precious freight, with the Stars and Stripes streaming in the wind, we felt a security and an indebtedness, which we now have all come to acknowledge. You went to New York, passed the capital of your country, and in a few days stood on the bloody fields of Antietam. There, in the old Ninth corps, under the faithful Burnside, you first realized war, and stood on the fated field of death. That was a sad day for the 16th Connecticut. Two hundred of your comrades, officers and men, had fallen, and when the shadows of evening closed on that historic day, your hearts, sad and broken, went up in thankfulness to God for your preservation. Well do I remember the sadness that settled over this city, as the news came across the wires that death had cruelly thinned and decimated your ranks, and that among the brave who had fallen were the noble Captains Manross, Drake, Brown, and Barber; but you had only time to bear out the noble two hundred to their sepulture, and to place some humble stone to mark their rest, before you were ordered to the fields of Fredericksburg.

There bravely you stood through three long days of battle. From there you moved to Suffolk, where your ranks were again broken, and the brave Capt. Tennant fell—he who was beloved at home and dear to the Hartford City Guard, of which he was a member, and who followed him with reversed arms and bowed hearts, to yonder church yard where now he sleeps. Peaceful will be his rest, and sacred his memory, for he died for country and humanity. Slowly the days went by as you were garrisoned at Portsmouth, and long were your marches from Plymouth to New Berne. You will never forget, but will tell over to your children through all the coming years of your lives, how after a long siege by the rebels at Plymouth, nearly your entire regiment was captured, and taken away to rebel prisons at Macon and Andersonville; how of the four hundred and four unfortunate captives who entered Andersonville prisons, more than half were borne out lifeless and cold. My blood chills when I remember that more than two hundred of this regiment were starved and murdered in Southern prisons—imaging more perfectly the hell of secession and the barbarism of Southern institutions and chivalry, than any other page of the war. But you have been preserved to come back to your homes, and to watch again the flag of your country waving over your native city. You well remember how beautiful it looked to you when first you saw its colors after the hour of your deliverance came, and you passed the rebel lines. That flag to-day symbolizes the greatest freedom and the most perfect nationality. "'Tis the banner of all the West," that of a nation now pronounced—

"The heir of all the ages in the foremost files of time."

But your work is done and your history is sealed. In the name of those for whom you fought and who have gathered here to do you honor it is my privilege to say a thousand times welcome home. Your thinned ranks, your torn colors, give convincing proof of your deeds of bravery The state will hang your tattered flag in yonder capitol, and claim with pride your history, and through all the future of the country, her sons will love, respect, and honor you as the brave soldiers who fought in the last great battle for freedom.

But our city has a special interest in your history. You have been led by the honored Col. Beach; and the name of your Lieutenant-Colonel, he who would not leave his regiment for the colonelcy of another, he who has been with you in the camp, in the battle, in the prison and on the march until now, than whom there is none better, or braver on all the veteran roll, his name, John H. Burnham, has long been with us as a household word. Think it not strange that the Hartford City Guard extends a warm and earnest welcome to you, for the names of Burnham, Cheney, Tennant, Pasco, Burke,

Lockwood, and Blakeslee, are all ours. As brothers we have watched your history during all these years of war. But while you are privileged to return all covered with honor and glory, and are to go to your homes, to be welcomed by mothers, wives, sisters, and lovers, noble patriotic women, in whose life there dwells the tenderest sentiment for you and country ever unlocked from the starry skies,—while you are to enjoy all this, I cannot forget the brave, devoted boys, comrades in all your trials until death, who will never, never come home again. They died, as you have fought, for country, for the restoration of law and order, for the complete emancipation of a race, for the eternal principle of liberty, and for the final solution of the great problem of self-government. They fell away from home and friends, and most of them rest in Southern graves, but though they fell thus, they died at their posts. History will keep fresh their memories, and write their names on more than granite shaft or marble column.—

After an eventful life and a noble death, they rest well.

"Sleep sweetly, tender hearts, in peace,
Sleep, holy spirits, blessed souls,
While the stars burn, the moons increase,
And the great ages onward roll."

The friends of the dead of your regiment are more than of the living, and my heart was sad as I saw the tears start in the eyes of the little child, the tender maiden and the mother with her little ones, as they looked in vain among your passing ranks for their friends. But they will never again watch their returning footsteps, or hear the sweet sound of their voices. No words of mine can heal their wounded hearts. I can only say they have the highest claim upon the nations' gratitude. The noble deeds of their martyred dead will ever live in the archives of the State, and their memories will be embalmed forever in the feelings of the American people.

Thrice welcome then, tried and faithful veterans of the republic. Go bear your honors and your trophies to your homes, and around your own hearths be as great and good as you have been in war.

Breakfast was provided for the regiment at the Trumbull House and United States Hotel, after which a short street parade was had, the City Guard and Colt's band furnishing escort. Arms were then stacked in the armory of the Guard, and the regiment was dismissed, the men being allowed to go to their homes as most of them live in this vicinity, and fathers, mothers, brothers, sisters, and wives were waiting to extend their welcome. Before leaving the armory Colonel Burnham read the following farewell order:

Headquarters 16th Conn. Vols., Infantry,

Hartford, Conn., 29th June, 1865.

General Orders No. 10.

Soldiers of the Sixteenth Connecticut—Glad as I am that the war is over and we are all to rejoin our families and friends again, I cannot repress a feeling of sadness at the thought of severing, perhaps forever, the ties that have bound us together for the last three years. Although a less amount of glory in the field has fallen to our lot than to some others, no regiment from the State has been subjected to so much suffering. Whatever you have been ordered to do, you have done promptly, cheerfully, and well; and whenever in future I am asked of what in all my life I am proudest, I shall always answer "that I belonged to the 16th Connecticut, in the Union army." Placed in charge of the regiment under circumstances that might have made my task a difficult one, it will always be a source of the highest gratification to me to remember that I received the generous support of all, and to the effective co-operation of every officer and the high character of the men, belongs entirely the credit of the fine soldierly appearance and superior military condition in every respect you so soon acquired, and have not failed to maintain to this day. For those gallant comrades who have laid down their lives on the battle-field and in the hospital, and for those heroic men who have endured so much more than death in Southern prisons, let us shed a silent tear and ever cherish their memories tenderly in our hearts.

You are about to go to your homes to enjoy in peace the blessings of the great and good government you have done your share to defend and preserve to our magnificent country, and I desire that each one of you should take with him my sincere thanks, for the cordial and hearty support you have given me, and the cheerful manner in which you have performed every duty required of you while I have the honor to be your commander. Wishing you every success that you can desire, and trusting you may find in your restored firesides all the comfort and happiness you have so nobly earned, I bid you an affectionate farewell.

John H. Burnham,

Lieutenant-Colonel 16th

Conn. Vols. Infantry.

On July 8th, the regiment assembled in Hartford, and were paid in full, except the officers who received their pay when returns were made and all property properly accounted for. Thus ended the organization known as the Sixteenth Regiment Connecticut Volunteers. Its record is a credit to the State and an honor to every man who has shared in its fortunes.